T0220310

Cambridge Elements ≡

Elements in the Philosophy of Science
edited by
Jacob Stegenga
University of Cambridge

BAYESIANISM AND SCIENTIFIC REASONING

Jonah N. Schupbach
University of Utah

CAMBRIDGE
UNIVERSITY PRESS

University Printing House, Cambridge CB2 8BS, United Kingdom

One Liberty Plaza, 20th Floor, New York, NY 10006, USA

477 Williamstown Road, Port Melbourne, VIC 3207, Australia

314–321, 3rd Floor, Plot 3, Splendor Forum, Jasola District Centre, New Delhi – 110025, India

103 Penang Road, #05–06/07, Visioncrest Commercial, Singapore 238467

Cambridge University Press is part of the University of Cambridge.

It furthers the University's mission by disseminating knowledge in the pursuit of education, learning, and research at the highest international levels of excellence.

www.cambridge.org
Information on this title: www.cambridge.org/9781108714013
DOI: 10.1017/9781108657563

First published 2022

A catalogue record for this publication is available from the British Library.

ISBN 978-1-108-71401-3 Paperback
ISSN 2517-7273 (online)
ISSN 2517-7265 (print)

Bayesianism and Scientific Reasoning

Elements in the Philosophy of Science

DOI: 10.1017/9781108657563
First published online: January 2022

Jonah N. Schupbach
University of Utah
Author for correspondence: Jonah N. Schupbach,
jonah.n.schupbach@utah.edu

Abstract: This Element explores the Bayesian approach to the logic and epistemology of scientific reasoning. Section 1 introduces the probability calculus as an appealing generalization of classical logic for uncertain reasoning. Section 2 explores some of the vast terrain of Bayesian epistemology. Three epistemological postulates suggested by Thomas Bayes in his seminal work guide the exploration. This section discusses modern developments and defenses of these postulates as well as some important criticisms and complications that lie in wait for the Bayesian epistemologist. Section 3 applies the formal tools and principles of the first two sections to a handful of topics in the epistemology of scientific reasoning: confirmation, explanatory reasoning, evidential diversity and robustness analysis, hypothesis competition, and Ockham's Razor.

Keywords: Bayesianism, explanatory reasoning, formal epistemology, inductive logic, scientific reasoning

ISBNs: 9781108714013 (PB), 9781108657563 (OC)
ISSNs: 2517-7273 (online), 2517-7265 (print)

Contents

Introduction

This Element introduces and critically explores the Bayesian approach to the logic and epistemology of scientific reasoning. This approach is distinguished by its conceptual *medium* and formal *method*. Bayesians paint in the medium of what has variously been called "degree of belief," "credence," "confidence," and so on – which is to say that Bayesian accounts essentially involve a gradational notion of the doxastic attitude that agents have toward propositions. Some common terminology notwithstanding (e.g., "partial belief" and "degree of belief"), this concept is not a straightforward, gradational version of *belief*, which would be referring to something like proportion of outright belief. Such a notion would "max out" in the case of belief simpliciter, but Bayesians tend to be very quick in dismissing the identification of maximal degree of belief, credence, and so on with qualitative belief (Maher, 1993; Leitgeb, 2013; Buchak, 2013). Much more commonly, the Bayesian's gradational medium is described as maxing out in the special case of *certainty* (e.g., Ramsey 1926, §3; Jaynes 2003; Jeffrey 2004; Leitgeb 2013; Sprenger and Hartmann 2019, p. 26). Certainty is a doxastic, propositional attitude of agents generalizing naturally to less extreme attitudes of *uncertainty* or *confidence*. Accordingly, in an attempt to avoid confusion in this Element, I use these latter terms when referring to the concept at the core of Bayesian accounts.[1]

Regarding method, Bayesians conduct their investigations using the mathematical tools of probability. This formal approach enables Bayesians to pursue their philosophical work with rigor and precision. The use of probability theory goes hand in hand with Bayesianism's focus on confidence; indeed, Bayesians view the probability calculus as an apt formal tool because of their emphasis on confidence and uncertainty. The bridge between medium and method here is the Bayesian's epistemic interpretation of probabilities as degrees of (more or less operationalized and/or idealized conceptions of) confidence.

Terms of art in probability theory – like "probable" and "likely" – are often used in everyday language to communicate epistemic judgments of uncertainty. However, the Bayesian interpretation is far from uncontroversial. Probability theory, mathematically speaking, is a branch of measure theory, complete with an axiomatization and consequent structure (Kolmogorov, 1933). While the epistemic interpretation of probability as a guide to rational confidence was there from the beginning of this math's development, much of this development was driven by more objective, physical, and aleatory concepts, applications,

---

[1] The use of "partial belief" and "degree of belief" to refer to Bayesianism's central notion can indeed lead to confusion and criticisms of Bayesianism that are rather too easy. For example, see Horgan's (2017, p. 236) "conceptual confusion" objection to Bayesianism.

and interpretations (see Hacking 2006, ch. 2). The mathematics of probability was at best developed only in part as an explication of *confidence*. Thus, a case needs to be made for the validity and usefulness of the Bayesian's interpretation.

This Element explores the Bayesian approach to logic and epistemology in three parts. Section 1 provides a primer on the elementary mathematics of probability, motivated and presented through a Bayesian lens. We develop probability theory as a conservative generalization of classical logic, which is more readily applicable to reasoning under uncertainty. Probability theory provides a *logic of consistency* for attitudes of confidence. That is, probability theory describes how an agent's confidences (at a particular time) ought to relate in order for them to be internally consistent. Additionally, this section discusses the Bayesian interpretation of probability as a measure of confidence, and it critically evaluates some of the common arguments presented for and against this interpretation.

Section 2 explores a number of contentious principles put forward and debated by Bayesians. Unlike the rules discussed in Section 1, these principles are more characteristically and recognizably epistemological. This is because they go beyond the logic of consistency in theorizing about how our confidences ought to be, not just related to each other, but sensitive to and constrained by our experience of the world. For direction in our exploration, we turn to the patron saint of Bayesian epistemology, the good Reverend Thomas Bayes, and his seminal "Essay Towards Solving a Problem in the Doctrine of Chances" (1763). Bayes suggests three epistemological principles in this work, each of which is still much discussed, developed, and debated by contemporary Bayesian epistemologists. We discuss these principles in some detail along with some corresponding criticisms and complications that lie in wait for the Bayesian epistemologist.

The final Section 3 displays the potential fruitfulness of the Bayesian approach for the study of scientific reasoning. We introduce just a handful of the many topics Bayesians have discussed in the epistemology of science: specifically, we focus on the epistemology of confirmation, explanatory reasoning, evidential diversity and robustness analysis, hypothesis competition, and Ockham's Razor. Via our discussions of these topics, we aim to show that our understanding of some important concepts, methods, and strategies commonly used in scientific practice can be improved by taking a Bayesian approach.

For the sake of keeping this work Element-length, I've often had to refrain from delving into interesting issues and obvious questions left standing by my presentations. At one level, this necessity has bothered me, since it opens me up to criticisms that I might have tried to preempt and limits me to simplistic overviews of certain topics. However, at another level, I think and hope that the result is more conducive to promoting further discussion (be it in classrooms or research settings), and indeed to inspiring a wider variety of future research on the relevant issues.

I first learned about Bayesianism through a series of courses I took from Timothy McGrew as an MA student at Western Michigan University (from 2004 to 2006). To this day, I owe Tim an enormous debt of gratitude, not only for introducing me to an enjoyable and fascinating field of study, but also for being such a patient, caring, and careful instructor. The "dartboard representations" I employ in Section 1 trace their roots to similar diagrams Tim developed and used when introducing me to the field. My deepest thanks additionally go out to several colleagues and students who willingly and graciously gave their time to read drafts of this work and provide me with feedback. Jonathan Livengood and Joshua Barthuly in particular each carefully read and provided substantial feedback on complete drafts of this Element; this work is much better because of their gracious help. Other readers who gave me invaluable feedback on portions of the text include Jean Berroa, Liam Egan, Samuel Fletcher, Konstantin Genin, David H. Glass, Qining (Tim) Guo, Robert Hartzell, Daniel Malinsky, Conor Mayo-Wilson, Lydia McGrew, Jacob Stegenga, Michael Titelbaum, Jon Williamson, and two anonymous reviewers for Cambridge University Press. The material for §3.3.2 and a large part of §3.4 was developed in collaboration with David H. Glass. My heartfelt gratitude also goes out to Abbot Silouan and all of the monks at the Monastery of the Holy Archangel Michael (Cañones, New Mexico) for providing me with gracious hospitality, good conversation, and the unimaginably peaceful environment of their guesthouse, where I wrote the bulk of this Element.

While working on this Element, I was supported by the Charles H. Monson Mid-Career Award administered by the University of Utah's Philosophy Department and by grant #61115 from the John Templeton Foundation (which I co-directed with David H. Glass). The opinions expressed in this Element are those of the author and do not necessarily reflect the views of either of these funding sources.

Finally, "thank you" is not enough when I try to imagine how I should respond to the support, love, and joy that I receive everyday from my wife and children. I don't deserve such blessings. Truth be told, my confidence is not very high that any one of them will ever read the entirety of this Element. Nonetheless, it wouldn't have happened were it not for them, and I dedicate it to them.

1 Probability Theory, a Logic of Consistency

Bayesianism is characterized by its focus on the notion of confidence ("degree of belief," "credence," "partial belief," etc.) and its corresponding epistemic interpretation of probability theory.[2] Bayesians apply the probability calculus, so interpreted, to a wide variety of epistemological principles, concepts, and puzzles. This section introduces the mathematics of probability theory through a Bayesian lens. By first touching on some features of classical, deductive logic, we highlight the need to generalize this formal logic in order to deal directly with the inevitable uncertainties of scientific (and everyday) inferences. The probability calculus, when given a Bayesian interpretation, provides a particularly appealing generalization, resulting in a compelling logic of consistency for the confidences of uncertain agents. We finish this section by considering some arguments for and against this way of construing probability theory.

1.1 Logic and Uncertainty

Logic is the science of inference. This discipline systematically studies the relation by which conclusions follow from premises. As such, while logic is fundamental to epistemology, the two disciplines are distinct. Epistemology would surely be lacking if it gave no role to inference in the theory of knowledge, but there are other factors besides inference that play crucial epistemological roles.[3] Logic is also distinct from another science of inference, the psychology of inference, which observes, predicts, and models the inferential behavior of humans actually drawing conclusions from premises.[4] Logic, by contrast, is more of a theoretical science. It theorizes about the existence and nature of general principles relating conclusions and premises in such a way as to legitimize, guide, and regulate such behavior. These principles, which the renowned logician George Boole (1854) called the "laws of thought," concern the fundamental nature of consequence, or what follows from what.

---

[2] Bayesianism is often additionally characterized as being committed to **Bayes's Rule**, a principle legislating how confidences (explicated as probabilities) should evolve over a diachronic process of learning new information. We postpone our discussion of this principle until Section 2.

[3] There are other conceptions of logic, some of which encompass epistemology. For example, Ramsey (1926, p. 87) views logic as consisting of two parts, "formal logic" or the explicative logic of consistency, and "inductive logic" or the ampliative logic of discovery and truth. I accept Ramsey's distinction but am trying to keep terminology tidy by reserving "logic" for Ramsey's notion of formal logic. What Ramsey calls "inductive logic" is part of epistemology as I characterize it – cf. Jeffreys's (1939, p. 1) remark: "The theory of learning in general is the branch of logic known as epistemology."

[4] The wisdom of distinguishing the logical from the psychological may be questioned, however. Kimhi (2018), for example, argues that the sundering of the two is a recent, Fregean development that has had detrimental philosophical effects.

Logic is also sometimes thought of as the study of *valid* inference, in which the truth of a target conclusion follows *inescapably* from the collective truth of a set of corresponding premises (i.e., such that the conclusion cannot possibly be false if the premises are all true).[5] But while this seems a fair characterization of classical, deductive logic, the science-of-inference and study-of-valid-inference conceptions of logic may come apart for at least two related reasons. First, some inferences are logical in the sense that their conclusions follow from corresponding premises, even though they don't follow as a matter of necessity (i.e., even though the inferences aren't valid). Second, it may be useful to examine a notion of inference that relates propositions with respect to something other than their truth values. Both of these considerations underlie the approach to *inductive* logic that we take in this Element.

At the heart of both considerations is the attempt to accommodate the ubiquitous presence of uncertainty in scientific reasoning (and human reasoning more generally). Apart from some very special contexts in which truth values may legitimately be directly known with certainty, or stipulated, or constructed, and so on, reasoners are plagued by uncertainty. They simply do not have omniscient access to the truth values of propositions. What they have instead are inferior surrogate attitudes toward propositions, including attitudes of confidence and uncertainty.

Knowing how propositions relate to each other in terms of their truth values can still be very useful information for an uncertain agent. However, an alternative, inductive approach to logic seeks to shed light more directly on how propositions relate to each other in terms of the uncertain attitudes agents actually have toward them.[6] Instead of studying what truths follow from other truths, this approach studies what confidences, expectations, uncertainties, and the like follow from other such attitudes. And on this approach, it may be that some *invalid* inferences are nonetheless good insofar as (and possibly to the extent that) their conclusions are made more nearly certain by their corresponding premises. We seek a logic better suited for mortals,

---

[5] While some writers use the term "valid" to apply to good inductive arguments (e.g., Priest 2006; Sprenger and Hartmann 2019), in this Element, I reserve the term strictly for the notion of deductive validity.

[6] Alternatively, one might think of inductive logic as relating propositions in terms of their truth values, while generalizing the notion of *consequence* itself. From this perspective, inductive logic is still about what truths follow from other truths; however, the salient notion of consequence is generalized to allow for degrees of "partial entailment" and/or non-monotonic inference. Bolzano's (1837) notion of "relative satisfiability" is an early example of this perspective (see Howson 2011, §2). Other examples include Wittgenstein's interpretation of probability in the *Tractatus* (1922, §5.15) (see Williamson 2017, §1.1), "logical interpretations" of probability in terms of "partial entailment" (Keynes, 1921; Jeffreys, 1939; Carnap, 1962), and Priest's (2006, pp. 189–190) account of "inductive validity" using a non-monotonic logic.

a logic of uncertain inference for agents like us who don't have omniscient access to truth values.

Scientific practice provides any number of instances of compelling arguments in contexts of uncertainty. For example, in *On the Heavens* Aristotle builds a case from multiple lines of evidence for accepting the sphericity of the earth. His arguments provide a compelling interplay of observational evidence and inference.[7] He cites the evidence of the earth's shadow's circular shape as observed during a lunar eclipse. He then infers from this evidence and from his (accurate) understanding of lunar eclipses that the earth is a sphere, since that shape would account for this observation, given that theory. Similarly, Aristotle records observations of an "alteration of the horizon" relative to the fixed stars effected by northward or southward travel. He then infers from this evidence that the earth is a sphere, since again that shape would account for differences in the visible stars depending on an observer's latitude.[8] Distilled into a more organized form, the arguments are as follows:

Argument 1:
A1. If the earth is spherical, it would cast a circular shadow.
A2. The earth casts a circular shadow.
  C. Thus, the earth is spherical.

Argument 2:
B1. If the earth is spherical, northward and southward travel would alter the range of visible stars.
B2. Northward and southward travel alters the range of visible stars.
  C. Thus, the earth is spherical.

Both of these arguments are still used today and considered to provide powerful reasons for accepting the earth's sphericity. And they both exemplify an

---

[7] It's a longstanding and stubbornly persisting myth (still taught in schools, despite repeated corrections) that Christopher Columbus courageously embarked on his 1492 journey in the face of fears that his ship would fall off the edge of a flat earth. The truth is that virtually *no* educated Western European at the time of Columbus shared in such fears. Not only did the ancient Greeks discover the earth's shape, but they also made impressively accurate calculations of its size. This enduring myth was invented in the 1820s by the American writer, Washington Irving, and propagated in his *History of the Life and Voyages of Christopher Columbus* (Lindberg, 2007, p. 161). For more on the invention of this flat earth myth, see Russell (1991) and Garwood (2008).

[8] Aristotle's own presentation of these arguments runs as follows: "How else would eclipses of the moon show segments shaped as we see them? [...I]n eclipses the outline is always curved; and, since it is the interposition of the earth that makes the eclipse, the form of this line will be caused by the form of the earth's surface, which is therefore spherical. Again, our observations of the stars make it evident, not only that the earth is circular, but also that it is a circle of no great size. For quite a small change of position to south or north causes a manifest alteration of the horizon. There is much change, I mean, in the stars which are overhead, and the stars seen are different, as one moves northward or southward." – *De Caelo* 297^{b}24-34, as translated in Barnes (1984).

extremely common style of scientific reasoning: we confirm a hypothesis by observing evidence that we expect to find if the hypothesis is true. But uncertainty manifests itself in these arguments in both of the aforementioned ways. First, the conclusion doesn't follow inescapably in either case; that is, the inference in both cases is invalid and thus uncertain. In fact, as any astute intro to logic student will quickly recognize, both arguments – if construed as attempts at deductive argumentation – commit the ostensibly pernicious, *elementary* fallacy of affirming the consequent! The earth could have some nonspherical shape (e.g., a flat disc) that still casts a circular shadow on the moon because of its orientation with respect to the Sun; and north/southward travel would result in a change of horizon with respect to the fixed stars if the earth were, for example, a properly oriented cylinder instead of a sphere. So neither argument's premises force it to be true that the earth is a sphere. A classically deductive approach, by focusing exclusively on valid inference, will thereby neglect any sense in which these invalid arguments are nonetheless good.

Second, reasoners aren't handed the truth values of these premises. For example, in **Argument 1**, A2 is particularly questionable since the shadow observed during lunar eclipses is, at any one time, partial with fuzzy boundaries. A1 is also dubious; even if the earth were a perfect sphere (which, of course, it's not), we wouldn't expect it to cast a *perfectly* circular shadow onto the moon insofar as the surface of the moon is itself curved and includes significant elevation changes (mountains and craters). At best, we are highly confident in the approximate truth of both premises, but classical, deductive logic tells us nothing about what to do with such attitudes and what they may or may not imply about the attitude we should take toward conclusion C.

The motivation behind the approach taken here is to develop a logic that can make sense of uncertain inferences like Aristotle's. If our logic is going to make room for such inferences, then it must allow for a sense in which deductively invalid arguments can be cogent. And if our logic is to guide agents like us in reasoning similarly, it should instruct us with respect to the uncertain attitudes we are working with in such cases.

1.2 From Deductive Logic to Probability Theory

Here, we introduce the probability calculus as a promising inductive logic. While we ultimately want to move beyond a logic that relates propositions in terms of truth values, we will presently find reasons to think that the formal semantics of deductive logic should be retained as a limiting case of our inductive logic. Accordingly, it will prove useful for us to begin with a quick review of the formal language of classical, *propositional* logic.

Once we've defined the vocabulary and grammar of propositional logic, we have all we need to distinguish well-formed (grammatical) from ill-formed (ungrammatical) parts of a formal-logical language. In this Element, we'll use italicized, capital letters like E, H, K, P, Q, R, and S as the basic atoms of our formal language. They count as grammatical by themselves. More complex statements of propositional logic can be formulated using standard "connectives," like $\neg$, $\wedge$, and $\vee$. These don't just connect to atomic formulae (capital letters) but may be used to connect any grammatical statements of the formal language, according to the following recursive definition (here and throughout, we use lowercase Greek letters as metavariables standing in for *any* grammatical formula of the language):

Grammatical formulae for propositional logic:
- Capital letters (possibly with subscripts) are grammatical formulae;
- If ϕ is a grammatical formula, then so is $\neg\phi$;
- If ϕ and ψ are grammatical formulae, then so are $(\phi \vee \psi)$ and $(\phi \wedge \psi)$;[9]
- Only formulae that can be shown to be grammatical by the above conditions are grammatical.

These capital letters and connectives are supposed to mean something to us; they're specifically meant to be more exact versions of familiar components of our natural language. For example, $\neg$ is supposed to be like the English word "not," $\wedge$ like "and," and $\vee$ like the English inclusive "or" ("inclusive" meaning that the "or" statement doesn't rule out the possibility of both statements it connects being true). But it's important to note that, at *this* point in the development of the logical language, a grammatical formula like "$(P \wedge (\neg Q \vee R))$" means no more than complete gibberish, like "$((\vee \wedge *PD$". In order to make our grammatical statements mean something, we need to go one more crucial step and specify their semantics.

Classical logic's reliance on truth values becomes apparent (and incredibly helpful) at this point. The formal semantics of propositional logic is straightforwardly "truth-functional," meaning that its connectives' meanings are specified precisely by articulating the truth values they output as a function of all possible combinations of truth values they take in. The connective $\neg$ denotes *negation*, the truth-functional operation resulting in a true statement if and only if

---

[9.] The connective for the material conditional $\rightarrow$ is conspicuously absent from this list. In the rare cases where this connective makes an appearance, we'll think of it as part of our non-primitive vocabulary, $\ulcorner(\phi \rightarrow \psi)\urcorner$ being short for $\ulcorner\neg(\phi \wedge \neg\psi)\urcorner$. In fact, $\rightarrow$ will not play a substantial role in our discussions.

Though all grammatical disjunctions and conjunctions officially are enclosed in parentheses, we will follow a standard practice and drop *outermost* parentheses – i.e., any parentheses that have the entire remainder of the formulae within their scope.

the statement negated is false. The ∨ denotes *disjunction*, the truth-functional operation resulting in a false statement if and only if both of the connected "disjuncts" are false. The ∧ denotes *conjunction*, the truth-functional operation resulting in a true statement if and only if both of the connected "conjuncts" are true. Articulating this formal semantics in terms of the standard truth tables (and again using lowercase Greek letters as metavariables standing in for any grammatical formula of the language), we have:

ϕ	$\neg\phi$
T	F
F	T

ϕ	ψ	$\phi \vee \psi$
T	T	T
T	F	T
F	T	T
F	F	F

ϕ	ψ	$\phi \wedge \psi$
T	T	T
T	F	F
F	T	F
F	F	F

From the viewpoint of inductive logic, there is nothing amiss thus far. Defining these connectives by specifying their truth-functional roles makes sense even if you deny that reasoners have direct access to such truth values. Nonetheless, while classical logic may offer an appropriate truth-functional semantics, that formal semantics falls short of providing us with an inductive logic of confidences.

In working toward such a logic, the following interesting point is crucially important. Classical logic's truth-functional semantics doubles as a plausible *certainty*-functional semantics. That is, we can interpret "T" as "certainty of truth" as opposed to truth simpliciter and "F" as "certainty of falsehood" as opposed to falsehood simpliciter, and all of the tables still turn out right. For example, this reading of the negation sensibly associates certainty that $\neg\phi$ is true [false] with certainty that ϕ is false [true]. This shift in interpretation is substantial; instead of trading directly in truth values, it trades in attitudes that we might have toward propositions, albeit extreme attitudes that we may have only rarely. To mark the shift in interpretation, let's rewrite the above tables using different values; we'll use "1" for certainty of truth and "0" for certainty of falsity:

ϕ	$\neg\phi$
1	0
0	1

ϕ	ψ	$\phi \vee \psi$
1	1	1
1	0	1
0	1	1
0	0	0

ϕ	ψ	$\phi \wedge \psi$
1	1	1
1	0	0
0	1	0
0	0	0

Using numerical values marks our shift in interpretation, but this move also – regardless of interpretation – allows us to represent deductive logic's formal semantics algebraically. Let *Val* be the "valuation function" that assigns to classical logic's grammatical formulae a member of the set $\{0, 1\}$.

The first and third tables are then straightforwardly summarized in terms of *Val* as follows:

- $Val(\neg\phi) = 1 - Val(\phi)$.
- $Val(\phi \wedge \psi) = Val(\phi) \times Val(\psi)$.

The case of $\vee$ is somewhat less obvious, and it proves instructive to flounder a bit trying to represent its table algebraically. Were it not for the first line of $\vee$'s table, the following straightforward sum operation would do the trick: $Val(\phi \vee \psi) = Val(\phi) + Val(\psi)$. In other words, in the special case where the first line can be ignored, this simple sum operation would be correct. But we're right to ignore the first line as a genuine possibility exactly when ϕ and ψ cannot possibly be jointly true – that is, when they are "mutually exclusive." This is worth emphasizing:

- If ϕ and ψ are mutually exclusive, then $Val(\phi \vee \psi) = Val(\phi) + Val(\psi)$.

Of course, we still want a general algebraic representation for disjunction. The reason that the above rule doesn't work for the case when we are certain that both ϕ and ψ are true is because $Val(\phi \vee \psi)$ would equal two instead of the appropriate value of one according to the straightforward sum. To correct for this, we could just subtract out a function that takes value one when $Val(\phi) = Val(\psi) = 1$ and 0 otherwise. We already have such a function in $Val(\phi \wedge \psi)$! Thus, our general algebraic representation for $\vee$ is:

- $Val(\phi \vee \psi) = Val(\phi) + Val(\psi) - Val(\phi \wedge \psi)$.

To recap, we are seeking a logic that deals in confidences and uncertainties instead of dealing directly in truth-values. Nonetheless, we have found it useful to start with the truth-functional semantics for classical logic, since this semantics doubles as a plausible *certainty*-functional semantics. This makes intuitive sense since extreme cases in which we have certainty of the truth or falsity of some proposition correspond exactly with those cases where we at least take ourselves to be dealing directly with truth-values. The upshot is that our inductive logic should retain the above algebraic rules, at least as limiting case rules that hold in contexts of certainty (i.e., degenerate cases of uncertainty or extreme cases of confidence).

Probability theory can then be thought of as a *conservative* departure from classical logic in the sense that it preserves all of the above algebraic rules for the connectives – and indeed uses two of these rules in particular to ground all of its mathematics. Probability generalizes the bivalent, classical semantics, not by broadening its rules but by widening its evaluative range from the bivalent set $\{0, 1\}$ to the closed unit interval $[0, 1]$. Accordingly, let Pr be a function that assigns to each of propositional logic's grammatical formulae a real number between 0 and 1 inclusive. Now adopt, from the above, the following two rules for any grammatical formulae ϕ and ψ (now written in terms of Pr):

Negation. $Pr(\neg\phi) = 1 - Pr(\phi)$.
Additivity. For mutually exclusive ϕ and ψ, $Pr(\phi \vee \psi) = Pr(\phi) + Pr(\psi)$.

This basic setup already encompasses the entirety of probability theory! For example, the orthodox axiomatization of probability for finite applications (Kolmogorov, 1933, p. 2) effectively lays down three axioms for the probability calculus, **Additivity** plus the following two:

Nonnegativity. $Pr(\phi) \geq 0$.
Normality. For any tautology **T**, $Pr(\mathbf{T}) = 1$.

Both of these axioms are entailed by the setup above. **Nonnegativity** follows simply from our specification of Pr's range of $[0, 1]$, while **Normality** follows directly from **Negation** and **Additivity** (since ϕ and $\neg\phi$ are mutually exclusive):[10]

1. $Pr(\phi \vee \neg\phi) = Pr(\phi) + Pr(\neg\phi)$ (**Additivity**)
2. $Pr(\phi) + Pr(\neg\phi) = 1$ (**Negation**)
3. $Pr(\phi \vee \neg\phi) = 1$ (from 1 and 2).

At this point, note that we haven't given much of an argument for thinking that probability theory is *the* (or even *an*) appropriate formal logic of confidence. What we have done is motivated the idea that our inductive logic of confidence should agree with classical logic's semantics in limiting cases of certainty. We thus want inductive logic to generalize from classical logic, and we've argued that probability theory is such a straightforward, conservative generalization. At the end of this section, we'll consider some arguments for thinking that probability theory is the appropriate generalization of classical logic for capturing uncertain reasoning. For now, the idea stands merely as a natural conjecture worth considering and investigating further. But before we can proceed in more detail with the logical-philosophical investigation of this section, we must introduce more of the basic rules of the probability calculus.

---

[10] Strictly speaking, only instances of **Normality** for tautologies of the form $\phi \vee \neg\phi$ are shown to follow here. For the more general proof, we also need **Equivalence** – introduced in §1.3.3.

1.3 A Primer on Probability

1.3.1 Dartboard Representations

To motivate and exemplify further rules, it will be useful to have before us a simple stochastic setup. Let's throw some darts. Our game of darts is going to be pretty odd compared to the familiar barroom game. For one thing, we'll imagine that our dartboards are perfectly *square* instead of circular. They measure exactly 1 foot in width and height and thus have a total area of 1 ft^2. We'll consider different variations on this dartboard, with different ways of designing the spaces; in some cases, our spaces will even overlap one another. In all cases, we'll stipulate the following:

1. the boundaries between the spaces are infinitely thin,
2. the darts are infinitely sharp, and
3. any throw of a dart is perfectly haphazard.

We stipulate that the boundaries between regions are infinitely thin and the darts infinitely sharp so that we don't need to countenance the possibility of a dart's landing on boundary lines. What is meant by a dart throw being perfectly haphazard? Just that the dart is equally likely to land in any location on the board. Perhaps it's because the thrower is blindfolded and the dartboard is randomly placed in the room. We'll assume in the examples that follow that, by some miracle, the haphazardly thrown dart actually manages to land on the board. But where it is on the board is anybody's guess.

The dartboard pictured in Figure 1 is about as simple as they come. There are only two spaces, an A-space and a ∼A-space. The spaces are clearly not equally sized, the A-space having twice the area of the ∼A-space. Remembering our idealizing assumptions about the thinness of borders, sharpness of the dart, and haphazardness of throws,[11] along with our assumption that the dart in question does actually manage to hit the board, what is the probability of A: that the dart hits the A-space? Answering this question is as simple as calculating the proportion of the total dartboard area that the A-space takes up. That is, we need to calculate the rectangular area of the A-space, base times height, or ⅔ ft × 1 ft = ⅔ ft^2. The total area of the dartboard itself is 1 ft × 1 ft = 1 ft^2. And so the proportion of that total area taken up by the A-space is ⅔ ft^2 ÷ 1 ft^2 = ⅔, which allows us to conclude that $Pr(A) = 2/3$.

We should pause to appreciate at least two things before continuing our game of darts. First, one reason I chose to set up the dartboard with a total area of 1 ft^2 was to simplify our math. Notice that any proportional area relative to this entire dartboard is calculated as the area of the space in question divided

---

[11] For the most part, I'll leave these assumptions implicit from here on.

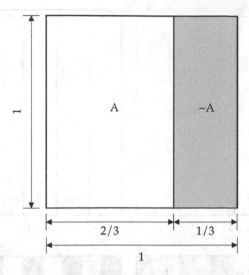

Figure 1 A 1 × 1 dartboard partitioned into two spaces.

by the board's total area of 1 ft². Dividing by one does nothing to change the value in question, and so we can more simply calculate any proportional area just as the area of the space in question. While I will usually make use of this shortcut from now on, it's important to remember that this works only when proportional areas are relative to a unit area (a total area of 1 ft²).

Second, notice that proportional areas are *dimensionless*, as are proportions more generally. At the end of our calculation of proportional area above, we divided ⅔ ft² by 1 ft². The units thus cancel out and we're left with the simple proportion of ⅔ (*not* ⅔ ft²). This should tip us off that our calculation was not only insensitive to the units we chose for measuring base, height, and area, but also that our probability problem ultimately had much more to do with proportions than with areas. We didn't calculate the area of the A-space because of any *analytic* connection between area and probability but rather because that gave us a means to calculate the *proportion* of possibilities in which the dart lands on an A-space.

This suggests an interpretation of our "dartboards" that applies far beyond the barroom. Instead of representing the possible landing sites of an imagined, haphazardly thrown dart, we might use the same setup to model the possible outcomes of any game, experiment, or general event. The probability of any event would then be represented as the *proportional* area of the space in which the event in question occurs within that complete possibility space. (Figure 2 gives some examples of "dartboard representations" of other games.) We may even try to model all possible states of affairs in the universe in a square! The probability of any proposition could then be represented as the proportional area of the space of states of affairs in which the proposition in question is true.

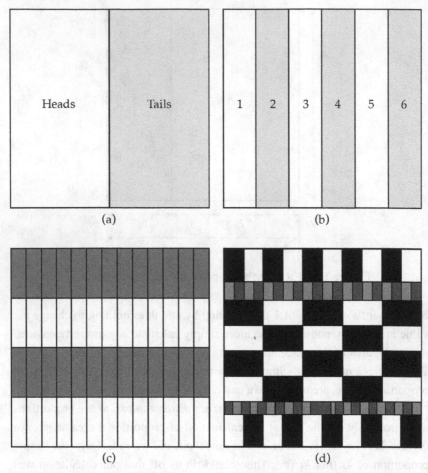

Figure 2 Dartboard representations of (a) the flip of a fair coin, (b) the roll of a fair die, (c) a random draw from a standard 52-card deck, and (d) perfectly haphazard hits on a standard dartboard. The smallest rectangle in (d), for example, has the same proportional area to the unit square as the inner bull of the bullseye has to the area of a standard dartboard; thus, the chance that a haphazardly thrown dart lands in the inner bull on a standard board is represented here by that rectangular area.

The catch is that this makes sense only to the extent that there is a meaningful sense in which the proverbial dart is just as likely to land in any location on the board.

1.3.2 Revisiting the Basic Rules

Let's go back to our simple dartboard (Figure 1) and demonstrate the rules of probability that we've already discussed. We've seen that $Pr(A) = 2/3$, but what is the probability of $\neg A$: that the dart does not hit an A-space? This probability may be calculated in two obvious ways. We could calculate the proportional area of the region outside the A-space (i.e., of the $\sim$A-space): ($\frac{1}{3}$ ft $\times$ 1 ft) $\div$ 1 ft^2 = $\frac{1}{3}$. But we could just as easily have subtracted the proportional area of the A-space from the unit area to see how much of the total unit area falls outside the A-space: $Pr(\neg A) = 1 - Pr(A) = 1 - \frac{2}{3} = \frac{1}{3}$. This latter procedure exemplifies **Negation**, which shows up diagrammatically as the idea that the area falling outside a space can be calculated by subtracting that space's area from the unit area of one.

Additivity applies to propositions that are mutually exclusive (i.e., to propositions that cannot be true conjointly). Propositions A and $\neg A$ are like this. In Figure 1, the A- and $\sim$A-spaces do not overlap, revealing that there is no possible outcome in which the thrown dart simultaneously hits them both. But what is the probability that a dart hits the A-space *or* the $\sim$A-space? **Additivity** requires this probability to be equal to the sum of their individual probabilities: $Pr(A \lor \neg A) = Pr(A) + Pr(\neg A)$. In the dartboard representation, this corresponds with the obvious fact that the total proportional area of two (or more) nonoverlapping spaces is equal to the sum of the spaces' individual proportional areas. The total proportional area of the A- and $\sim$A-spaces is found by adding their individual proportional areas together: ($\frac{2}{3} + \frac{1}{3}$) = 1. Of course, the total proportional area of two disjoint spaces need not equal the total unit area; for example, in Figure 2(b), the probability of rolling either a 1 or a 3 with a fair die corresponds to the total proportional area of the mutually exclusive 1- and 3-spaces: ($1 \times \frac{1}{6}$) + ($1 \times \frac{1}{6}$) = $\frac{1}{3}$.

Normality is similarly demonstrated. Any tautology is, by definition, true in all possible scenarios. For example, the tautology $A \lor \neg A$ is the logically true proposition that the dart either hits an A-space or not. All possible outcomes in the square dartboard satisfy this condition (also, since we're assuming that darts do successfully hit the board, there are no possible outcomes outside the square). Thus, the proportional area on the dartboard corresponding to $Pr(A \lor \neg A)$ is that of the entire square itself, or one. **Normality** is diagrammatically rendered as the association of tautologies (or sure events) with the entire unit area, the space of all possible outcomes. Finally, **Nonnegativity** refers to the diagrammatic constraint of there being no such thing as negative proportional areas.

1.3.3 Other Rules of Probability

A host of additional, powerful rules become evident if we consider a somewhat more complicated dartboard. The dartboard shown in Figure 3(a) includes two *overlapping* spaces. Their overlapping region is the bottom left $3/4 \times 1/3$ rectangle. Thus, the probability that a dart lands on an A- *and* a B-space (i.e., the "joint probability" of $A \land B$) is the proportional area of this "A&B-region," or $3/12 = 1/4$.

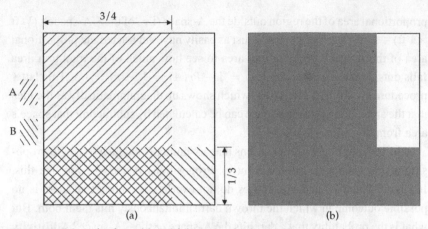

Figure 3 (a) A dartboard in which the A-space overlaps with the B-space. (b) The total combined area of these spaces.

Now we throw a dart and it lands on the board. First question: what is the probability that it lands on either an A- or a B-space ($A \vee B$)? We don't have a rule yet providing the answer to this question. **Additivity** comes close to giving us an answer, but it pertains only to situations in which the disjuncts in question are mutually exclusive (i.e., situations in which the corresponding areas don't overlap). Nonetheless, we can determine the answer in terms of dartboard areas. What we want is the proportional area of both spaces combined – the shaded region shown in Figure 3(b). If we foolishly applied **Additivity** here, we would think that the answer is simply the sum of the spaces' proportional areas $Pr(A) + Pr(B) = (3/4 \times 1) + (1 \times 1/3) = 13/12$! Something has clearly gone wrong, since probabilities max out at one.

What has gone wrong is that, by adding the A-space's area to the B-space's area, we've unwittingly double-counted their intersecting region. To correct for this, we need to subtract that region's area out once. Thus, the correct calculation is this: $Pr(A \vee B) = Pr(A) + Pr(B) - Pr(A \wedge B) = (3/4 \times 1) + (1 \times 1/3) - (3/4 \times 1/3) = 5/6$. Thinking about proportional areas has now taught us another rule for calculating the probability of a disjunction. Moreover, unlike **Additivity**, this rule is perfectly general, applying regardless of whether disjuncts are mutually exclusive:

General Additivity. $Pr(\phi \vee \psi) = Pr(\phi) + Pr(\psi) - Pr(\phi \wedge \psi)$.

Note that **General Additivity** subtracts out $Pr(\phi \wedge \psi)$ as a correction factor for cases in which the disjuncts are not mutually exclusive. And it incorporates **Additivity** as a special case, since this correction factor zeroes out ($Pr(\phi \wedge \psi) = 0$) in cases of mutual exclusivity.

It's worth taking a moment here to notice the parallel treatment of ∨ between classical logic and probability theory. Recall from §1.2 that the former includes both of the following rules governing its bivalent valuations:

- If ϕ and ψ are mutually exclusive, then $Val(\phi \vee \psi) = Val(\phi) + Val(\psi)$.
- $Val(\phi \vee \psi) = Val(\phi) + Val(\psi) - Val(\phi \wedge \psi)$.

Both of these rules transfer seamlessly over to probability theory. Thus, we're reminded that probability provides a conservative generalization of classical logic, preserving the algebraic rules governing the connectives, while applying them to assign an infinitely richer range of values.

Returning to the dartboard, let's consider a different sort of question: *Given that a dart lands in the A-space,* what is the probability that it lands in the B-space? The "given that" phrase in this question has us consider a narrower range of possibilities, excluding any dartboard locations located outside the A-space. That is, the A-space now delimits the possible locations of the dart (Figure 4(a)), and our question amounts to asking what the proportional area is of the B-space relative to this new possibility space. In other words, how much of the A-space does the B-space take up? This proportional area is: $(3/4 \times 1/3) \div (3/4 \times 1) = 1/3$. In probability terms, this amounts to dividing the joint probability of $A \wedge B$ by the probability of A. The result is called a "conditional probability," and it is denoted by listing the *given* information to the right of a new "|" symbol:[12] $Pr(B|A) = Pr(A \wedge B)/Pr(A) = 1/4 \div 3/4 = 1/3$. We've just uncovered a rule for calculating conditional probabilities:[13]

Conditional Probability.   If $Pr(\psi) > 0$, $Pr(\phi|\psi) = \dfrac{Pr(\phi \wedge \psi)}{Pr(\psi)}$.

Using this rule, we can easily calculate the corresponding probability that a dart lands on the A-space, given that it lands on the B-space:

$$Pr(A|B) = \frac{Pr(A \wedge B)}{Pr(B)} = \frac{1/4}{1/3} = 3/4.$$

---

[12] One might wonder why we need a new symbol for denoting "given that." Isn't the conditional probability of "ϕ, given that ψ" captured by the probability of "If ψ, then ϕ"? If so, perhaps we can use classical logic's material conditional to explicate this phrase. Lewis (1976) famously shows, however, that the identification of conditional probabilities with probabilities of conditionals (be they material or indicative conditionals) trivializes one's probability measure, rendering it completely useless.

[13] Most introductions take this rule of probability as a *definition* or conceptual analysis of conditional probability. One awkward consequence of this move is that it immediately rules out the possibility of there being meaningful probabilities of propositions conditional on propositions with probability zero (Hájek, 2003). How best to respond to such concerns, and how best to define conditional probability, are fascinating and complicated issues (Hájek, 2009; Easwaran, 2016, 2019) that this Element unfortunately must sidestep.

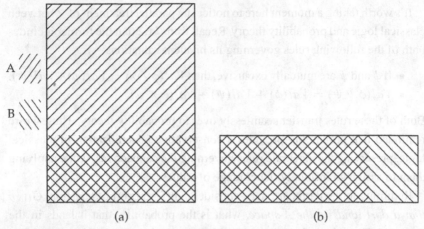

(a)   (b)

Figure 4 (a) The range of possibilities, given that the dart lands on the A-space. (b) The range of possibilities, given that the dart lands on the B-space.

A comparison of Figures 4(a) and (b) reveals how conditional probabilities adjust the appropriate range of possibilities via their specifications of *given* information.

In the above example, the probability that the dart is on a B-space interestingly remains the same *regardless of whether we are given that it landed on an A-space*; formally, $Pr(B) = Pr(B|A) = Pr(B|\neg A) = 1/3$. (The reader should, by now, be able to confirm that $Pr(B|\neg A) = 1/3$.) Since this is true, we say that B is probabilistically **independent** of A. Independence is a symmetric relation: B is independent of A if and only if A is independent of B.[14] Thus, we may infer (and leave it to the reader to verify) that A is also probabilistically independent of B in this same example: $Pr(A) = Pr(A|B) = Pr(A|\neg B)$.

There are many cases in which propositions are not probabilistically independent of one another. For example, let's alter our dartboard to look like Figure 5(a). While the proportional area of the A-space and the A&B-region has not changed, that of the total B-space has increased. As shown in Figure 5(b), its proportional area can be calculated as $3/4 \times 1/3 + 1/4 \times 2/3 = 5/12$. By contrast, the probability of a dart landing in the B-space, given that it lands in the A-space remains the same: $Pr(B|A) = 1/3$. Since $Pr(B) = 5/12 > 1/3 = Pr(B|A)$,

---

[14] This symmetry breaks down if either of the propositions in question has extreme probability (of 0 or 1). One can avoid this complication in either of two ways: 1) Define independence rather via the condition $Pr(\phi \wedge \psi) = Pr(\phi) \times Pr(\psi)$, which also holds symmetrically in extreme cases – see the **Conjunction (with Certainty)** and **Conjunction (with Independence)** rules below; or 2) stipulate that the notion of independence only pertains to propositions with non-extreme probabilities. We opt for the latter option here.

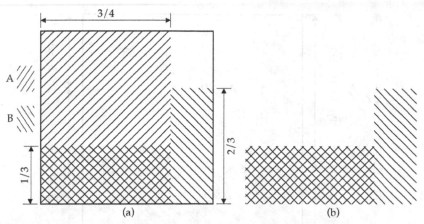

Figure 5 (a) A dartboard in which the A-space overlaps with the B-space, and for which *A* and *B* are no longer probabilistically independent. (b) The total B-space.

we say that *A* and *B* are negatively, probabilistically **dependent** on one another – "negatively" since the propositions "lower each others' probabilities." Had it been the case that $Pr(B) < Pr(B|A)$, we would say that *A* and *B* are *positively* dependent on one another (since *A* "raises *B*'s probability" in such a case). Just like independence, positive dependence and negative dependence are symmetric relations.

Like **Additivity, Conditional Probability** may be thought of as a special case rule, being applicable only when the denominator of its ratio term is positive. However, it straightforwardly corresponds to the following *general* rule of **Conjunction**.

Conjunction. $Pr(\phi \wedge \psi) = Pr(\phi) \times Pr(\psi|\phi)$.

We already effectively applied this rule when we calculated the probability of a dart's hitting both the A- and B-spaces, $Pr(A \wedge B)$, as the proportional area of the region in which the A- and B-spaces overlap. The base of this A&B region is equivalent to the base of the A-space, which, since the A-space has a height of one, is itself equal to $Pr(A)$. The height of this A&B region was equal to the height of the portion of the B-space that falls within the A-space; since this portion of the B-space and the A-space share the same width (and again since the A-space has a height of one), this height is equal to $Pr(B|A)$. Thus, when we calculated the area of the A&B region, we were instantiating the rule of **Conjunction:** $Pr(A \wedge B) = Pr(A) \times Pr(B|A)$.

Figure 6 reveals three other overlap regions we might define for our dartboard; besides the region in which the A- and B-spaces overlap, there is also, for

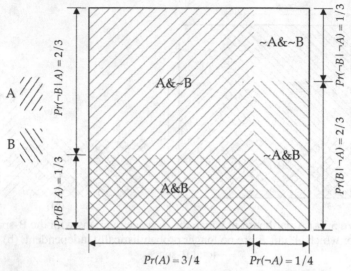

Figure 6 A dartboard representation of the joint probability distribution for *A* and *B*.

example, the A&~B region in which the A-space overlaps the area outside the B-space. This figure provides a diagrammatic justification for four applications of **Conjunction** that together specify the "joint probability distribution" for *A* and *B*:[15]

- $Pr(A \land B) = Pr(A) \times Pr(B|A) = 3/4 \times 1/3 = 1/4$
- $Pr(A \land \neg B) = Pr(A) \times Pr(\neg B|A) = 3/4 \times 2/3 = 1/2$
- $Pr(\neg A \land B) = Pr(\neg A) \times Pr(B|\neg A) = 1/4 \times 2/3 = 1/6$
- $Pr(\neg A \land \neg B) = Pr(\neg A) \times Pr(\neg B|\neg A) = 1/4 \times 1/3 = 1/12.$

At this stage, we can make a nice connection between classical, deductive consequence and probability theory. Notice that $A \land B$ is less probable (diagrammatically, has less proportional area) than either of the conjuncts entailed by it: $Pr(A \land B) < Pr(A)$ and $Pr(A \land B) < Pr(B)$. This holds true for the other conjunctions we just considered too, as the reader can verify. And a moment's reflection is all it takes to convince oneself that a conjunction cannot be more probable than either of its conjuncts. This follows generally from **Conjunction**, since $Pr(\phi \land \psi)$ is always equal to $Pr(\phi)$ reduced by some factor $Pr(\psi|\phi)$; it is only when this factor $Pr(\psi|\phi) = 1$ that a conjunction is as probable as one of its conjuncts. But a more general theorem follows if we simply think about

---

[15] Joint probability distributions are more carefully denoted relative to propositional *variables*, where such a variable has as its values the proposition or its negation; e.g., the propositional variable **A** takes values A or $\neg A$. When we talk about a joint probability distribution for two (or more) specific propositions, the reader should keep in mind that this is what we mean.

the nature of logical entailment. When some proposition ϕ deductively entails another proposition ψ (i.e., when $\phi \vDash \psi$), ψ would be true in at least all of the cases that ϕ would be true, and it may additionally be true in other cases. So, ψ has at least as much of a chance at being true as does ϕ:

Entailment. If $\phi \vDash \psi$, then $Pr(\psi|\phi) = 1$ and $Pr(\phi) \leq Pr(\psi)$.

The following extremely useful rule, pertaining to cases of logical equivalence (when two propositions entail each other), follows directly from **Entailment**:

Equivalence. If $\phi \dashv\vDash \psi$, then $Pr(\phi) = Pr(\psi)$.

As a quick example of this rule's usefulness, note that conjunctions $\phi \wedge \psi$ and $\psi \wedge \phi$ are equivalent to one another ($\phi \wedge \psi \dashv\vDash \psi \wedge \phi$). **Equivalence** thus implies $Pr(\phi \wedge \psi) = Pr(\psi \wedge \phi)$. Since, by **Conjunction**, $Pr(\phi \wedge \psi) = Pr(\phi) \times Pr(\psi|\phi)$ and $Pr(\psi \wedge \phi) = Pr(\psi) \times Pr(\phi|\psi)$, we may conclude that:

$$Pr(\phi \wedge \psi) = Pr(\phi) \times Pr(\psi|\phi) = Pr(\psi) \times Pr(\phi|\psi).$$

This is an even more general statement of **Conjunction**. In words, the probability of a conjunction equals the probability of *either* of its conjuncts multiplied by the probability of the other *given* the first.

In our brief study of the semantics of deductive logic (§1.2), we noticed the following principle for assigning 1s and 0s to conjunctions:

- $Val(\phi \wedge \psi) = Val(\phi) \times Val(\psi)$.

Unlike the classical rules for $\neg$ and $\vee$, this rule does not carry over without remainder to probability theory. **Conjunction** instead requires one of these propositions (either will do) to be evaluated *given* or *conditional on* the other. Nonetheless, there are two special cases in which **Conjunction** *is* identical to the deductive rule – the first of which shows that the classical, bivalent semantics for $\wedge$ is subsumed under probability theory:

Conjunction (with Certainty). If $Pr(\phi) \in \{0, 1\}$ or $Pr(\psi) \in \{0, 1\}$, then
$$Pr(\phi \wedge \psi) = Pr(\phi) \times Pr(\psi).$$
Conjunction (with Independence). If ϕ and ψ are probabilistically independent, then $Pr(\phi \wedge \psi) = Pr(\phi) \times Pr(\psi)$.

Consider the dartboard in Figure 5(a) again, and recall how we calculated the probability that a dart lands on the B-space. Referring to 6, we effectively summed up the proportional areas of the A&B-space and $\sim$A&B-space. That is, we calculated $Pr(B)$ as $Pr(A \wedge B) + Pr(\neg A \wedge B)$, which by **Conjunction** equals $Pr(A) \times Pr(B|A) + Pr(\neg A) \times Pr(B|\neg A)$. This is an instance of yet another important general rule:

Total Probability ($n = 2$). $Pr(\phi) = Pr(\phi|\psi) \times Pr(\psi) + Pr(\phi|\neg\psi) \times Pr(\neg\psi)$.

In order for this method to work, it was absolutely essential that A and $\neg A$ form a **partition**, that is, that (1) they be mutually exclusive and (2) they jointly exhaust the entire space of possibility (covering the entire dartboard area). In fact, we can generalize the rule of **Total Probability** ($n = 2$) to cover cases relating to partitions of any finite size:

Total Probability. For any partition of propositions $\{\psi_1, \psi_2, \ldots, \psi_n\}$,

$$Pr(\phi) = Pr(\phi|\psi_1)Pr(\psi_1) + Pr(\phi|\psi_2)Pr(\psi_2)$$
$$+ \ldots + Pr(\phi|\psi_n)Pr(\psi_n)$$
$$= \sum_{i=1}^{n} Pr(\phi|\psi_i)Pr(\psi_i).$$

To take an example in which $n > 2$, what would be the probability of hitting the E-space if we were playing with the dartboard shown in Figure 7? This dartboard is partitioned into five columns, H_1 through H_5; thus, the five propositions describing the dart landing on these respective spaces form a partition, $\{H_1, H_2, H_3, H_4, H_5\}$. The E-space is made up collectively of the lower portions of these columns (the total shaded area). The probability of hitting the E-space, $Pr(E)$, is equal to the area of the E-space. And this area can evidently be calculated simply as the sum of the areas (probabilities) of the various shaded regions: $Pr(E) = Pr(E \wedge H_1) + Pr(E \wedge H_2) + Pr(E \wedge H_3) + Pr(E \wedge H_4) + Pr(E \wedge H_5)$. Each of the areas corresponding to the individual summands is calculated simply by multiplying the corresponding base and height, resulting in the appropriate application of **Total Probability**: $Pr(E) = Pr(E|H_1)Pr(H_1) + Pr(E|H_2)Pr(H_2) + Pr(E|H_3)Pr(H_3) + Pr(E|H_4)Pr(H_4) + Pr(E|H_5)Pr(H_5)$.

Notice that the members of a partition must have probabilities summing to one. Accordingly, **Total Probability** may be thought of as stating that a proposition's probability can be calculated as a weighted average of its probabilities within each of an exhaustive set of disjoint possibilities. The weights are provided by the probabilities of the possibilities in question, which again must sum to one.

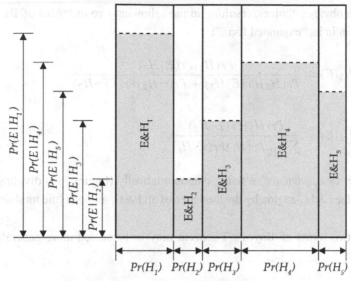

Figure 7 A dartboard representation of the total probability of E, calculated as the sum of the shaded proportional areas.

There is one remaining rule of probability that we need to introduce: **Bayes's Theorem**. This rule takes center stage in many formal epistemological studies of scientific reasoning (as we'll see in Section 3), but we'll stick to darts for now. Referring once more to Figure 7, we've already seen examples instructing us on how we would calculate the conditional probabilities of E given each H_i. These probabilities are easy to read off the dartboard representation, corresponding simply to the height of each E&H_i-region. Things are visually less straightforward if instead we ask for the probability that a dart lands on one of the H_i-spaces given that it lands on the E-space.

How, for example, do we calculate $Pr(H_2|E)$? The most detailed and helpful answer combines **Conditional Probability** with **Conjunction** and **Total Probability**. **Conditional Probability** gives us the following:

$$Pr(H_2|E) = \frac{Pr(H_2 \wedge E)}{Pr(E)}.$$

If we expand the numerator using **Conjunction**, we already have an instance of the simplest form of **Bayes's Theorem**:

$$Pr(H_2|E) = \frac{Pr(H_2)Pr(E|H_2)}{Pr(E)}.$$

A slightly more complicated (and typically more useful) version results if we then expand the denominator using **Total Probability**, relative to a partition that has H_2 as a member. The partitions $\{H_2, \neg H_2\}$ and $\{H_1, H_2, H_3, H_4, H_5\}$

are both obvious choices, resulting in the following two instances of **Bayes's Theorem** in its "expanded form":

$$Pr(H_2|E) = \frac{Pr(H_2)Pr(E|H_2)}{Pr(H_2)Pr(E|H_2) + Pr(\neg H_2)Pr(E|\neg H_2)},$$

$$Pr(H_2|E) = \frac{Pr(H_2)Pr(E|H_2)}{\sum_{i=1}^{5} Pr(H_i)Pr(E|H_i)}.$$

In either case, what we're doing diagrammatically (Figure 7) is dividing the area of the E&H_2-region by the total area of all E&H_i-regions (the total shaded area).

The two forms of **Bayes's Theorem** may be presented more generally as follows:

Bayes's Theorem (Simple Form). $Pr(\phi|\psi) = \dfrac{Pr(\phi)Pr(\psi|\phi)}{Pr(\psi)}.$

Bayes's Theorem (Expanded Form). For any member ϕ_i of the partition $\{\phi_1, \phi_2, \ldots, \phi_n\}$, $Pr(\phi_i|\psi) = \dfrac{Pr(\phi_i)Pr(\psi|\phi_i)}{\sum_{j=1}^{n} Pr(\phi_j)Pr(\psi|\phi_j)}.$

1.4 A Logic of Confidences

We are now ready for a deeper evaluation of the idea that the probability calculus provides a promising inductive logic. §1.2 introduced probability theory through the lens of a particular interpretation. We noticed that classical logic's truth-functional semantics doubles as a satisfactory certainty-functional semantics. We motivated the probability function *Pr* as a generalization of *Val*, governing attitudes of certainty of truth, certainty of falsehood, and everything in between. The "everything in between" includes attitudes of less than certainty, that is, uncertainties, or (non-extreme) confidences. In other words, the interpretation we've been assuming associates probabilities with confidences – which include extreme confidences, or certainties, and non-extreme confidences, or uncertainties.

But whose confidences are we talking about? The actual confidences had by real-life humans typically aren't as precise as probabilities. How can such precise entities as probabilities offer a useful logic of often imprecise epistemic attitudes like confidences? And anyway, why think that non-probabilistic confidences are illogical? The remainder of this section deals with these and related questions.

1.4.1 A Subject with No Object?

If probabilities are interpreted as confidences, whose confidences are they? The claim we want to consider is that probability theory constitutes a useful logic of consistency for confidences; that is, by clarifying how confidences ought to relate to one another (and/or what confidences follow from others), it provides normative constraints on how an agent's confidences *ought* to look. Accordingly, if probabilities qua degrees of confidence are properly thought of as being located in the psyche of epistemic agents, it is only with regard to fictional, ideal agents. Moreover, if the confidences represented by a probability measure don't show up in full in any actual human's mind, that is no mark against this theory, which after all claims to be normative, not descriptive.

Consider a parallel: classical logic may be useful as a logic of consistency for propositional beliefs, providing normative constraints on how one's beliefs ought to relate to one another. Maximal, deductively consistent sets of propositions may be thought to describe the beliefs of idealized logical agents. And while there may not be any actual human with a maximal, consistent set of beliefs, this is no mark against the deductive logic's claim to be laying down proper norms.

Still, there are various reasons why one might doubt that actual confidences could properly be legislated by comparing them to the probabilistic confidences of idealized agents. For one thing, the actual attitudes of confidence had by humans are rarely if ever sufficiently precise to be quantifiable with real values. Confidences are more typically comparative ("I'm more confident that the Cubs will win the division than that the Pirates will win it," or "I'm more confident than not that it will be sunny tomorrow") or qualitative ("I'm highly confident that Johnson will be reelected"). Indeed, it borders on the absurd to think of someone's announcing confidence .9228565... that it will be sunny tomorrow. But if actual confidences don't typically come with such precision, then why think they should be evaluated by such precise standards? The worry is that the so-called logical norms for confidence offered by probability theory would at best be purely theoretical, apt only for hypothetical superagents with epistemic states more precise and structured than those of humans. Horgan (2017, p. 241) voices this concern when he likens "Bayesian formal epistemology" to "past disciplines like alchemy and phlogiston theory; it is not about any real phenomena, and thus it also is not about any genuine norms that govern real phenomena."

In response to this concern, we can make two points. First, while it might border on the absurd to think that actual human confidences are typically quantifiably precise, there do exist some cases in which they plausibly seem to be.

Horgan himself offers some examples. A gameshow contestant knows that a prize is behind one of three doors (and nothing else relevant to the location of the prize). In such a case, it is not at all psychologically unrealistic to think that the contestant might have exactly equal confidence that the prize is behind any one of the three doors. But then quantitatively precise confidences are not the fictional entities that Horgan takes them to be. After all, this confidence judgment (along with the contestant's background knowledge that there is only one such prize hidden behind exactly one such door) can be expressed accurately with quantitative precision. Choose whatever conventions of quantitative explication you like; on the standard probabilistic, 0-to-1 scale of real values, the exact associated confidences are $\{1/3, 1/3, 1/3\}$.[16]

Saying that an agent has quantitatively precise confidences in such a case is crucially different from saying that the agent has quantified his or her own precise confidences using the probabilistic conventions, or any other conventions, or has done so at all! Rather, it suffices merely to show that the agent's confidences are quantifi*able*; that is, that they can be accurately represented with precise quantities.

The second point to make in response to this general concern is this: the precise, probabilistic confidences of idealized agents may still provide logical constraints on actual confidences, even when actual confidences are not quantifiable. This is because probabilistic confidences entail qualitative and comparative confidences. For example, probabilistic confidences of .9228565 that it will be sunny tomorrow and .0001193 that it will snow tomorrow imply quantitatively imprecise confidences, such as (most obviously) the comparative confidence that sun is more likely than snow tomorrow. More familiar and realistic attitudes of confidence can be read off from the idealized structure of agents with probabilistic confidences. Thus, a probabilistic inductive logic also speaks to cases in which actual agents have only comparative or qualitative confidences.

One might worry about the logical status of confidences that are less precise than their idealized counterparts. In the above example, if a logically idealized agent has these precise, probabilistic confidences that it will be sunny or

---

16 Of course, cases of quantifiably precise confidences need not be cases of equal confidence. Varying his example, Horgan imagines that the contestant is informed "that the location of the prize was based on the following, biased, randomizing procedure: after a fair die was tossed, the prize was placed behind door 1 if the die's up-face was either 1 or 2, behind door 2 if the up-face was 3, and behind door 3 if the up-face was 4, 5, or 6." If prompted, the contestant may affirm being twice as confident that the prize is behind door 1 than door 2, and three times as confident that it's behind door 3. Given the contestant's background knowledge, these realistic confidence judgments are quantitatively precise. On the standard probabilistic, 0-to-1 scale of real values, the exact associated quantitative confidences are $\{1/3, 1/6, 1/2\}$.

snowy tomorrow, is one somehow less logical for having only the comparative confidence that sun is more likely than snow? Here we come to another very important point. A normative model need not be (and I would surmise that no normative model should be) taken as prescriptive with respect to all of its features. Probability theory may be useful as a normative standard, representing idealized confidences of logical agents; but seeing it as such doesn't require thinking of all of the particulars of this model as aptly prescribed to actual agents. Indeed, there are reasons for introducing non-prescriptive idealizations into normative models (e.g., mathematical convenience). Not all idealizations involved in articulating a normative model are ideals (Colyvan, 2013). In our particular case, by assuming that idealized, logical agents have real-valued confidences across the gamut of propositions, we ensure that our normative model can speak to cases in which real agents actually have quantitatively precise confidences in any particular propositions. This idealization is thus not in place as a prescription for actual agents, but in order to ensure an appropriate level of generality or applicability for our logic.

To go back to the parallel case of classical logic, a maximal, deductively consistent set of propositions may be thought to describe the beliefs of an ideally logical agent. This is not to say that, when we evaluate actual epistemic agents by the lights of deductive logic, we condemn them for not having deductively closed, infinitely vast sets of beliefs – or for not considering all of the infinite number of propositions corresponding to the atoms of our formal system. These features of the idealized agent need not (and should not) be interpreted as prescriptions for actual agents. Instead, we evaluate those beliefs had by any actual agent by exploring the extent to which they might fit into an extended, idealized deductive system. The salient evaluative question is, "Could the beliefs that the agent actually brings to the table all be simultaneously believed by an ideal deductive agent?", not "Does this actual agent have all the same beliefs as an idealized deductive agent?".

As with deductive standards of rationality, so with inductive. A probability distribution may be thought to describe the confidences of an idealized agent, the agent thus having precise probabilistic confidences over a potentially infinite set of propositions. But this is not to say that, when we evaluate actual epistemic agents by the lights of this inductive logic, we should condemn them for not having quantifiable confidences over all (or even some) considered propositions – or for not considering in the first place all of the potentially infinite number of propositions in the domain. These features of the idealized probabilistic agent need not (and should not) be interpreted as prescriptions for actual agents. Instead, we evaluate the confidences had by any actual agent by exploring the extent to which they might fit into an extended, idealized inductive system. The salient evaluative question is, "Could the confidences that the agent actually brings to the table (be they quantifiable, qualitative, comparative, or otherwise) all be simultaneously had by an idealized probabilistic agent?", not "Does this actual agent have all of the same confidences as an idealized agent?".

1.4.2 Consistency of Confidences

Let's attempt a more careful articulation of what it takes for our probabilistic logic of consistency to validate an agent's confidences:

Consistency. An agent's confidences at a given time are consistent if and only if their formalizations are jointly entailed by at least one single probability distribution.

By "formalizations" we mean probabilistic explications of the agent's actual confidences. This is importantly different from saying that the confidences must be explicated as probabilities. If the agent has a comparative confidence of being more certain of ϕ than ψ, then the appropriate explication is similarly comparative: $Pr(\phi) > Pr(\psi)$. An alternative explication of this confidence that adds in numerical precision, like $Pr(\phi) = .4 > .28 = Pr(\psi)$, is manifestly inappropriate as it falsely portrays the agent's confidences as committed to a quantifiable level of precision, and this added precision might make all the difference with respect to **Consistency**. That is, such precise values may be inconsistent (according to **Consistency**) with other of the agent's confidences while the weaker inequality is not – for example, in the above case, if the agent is also more confident than not that ϕ.

The formalization $Pr(\phi) > Pr(\psi)$ mentions a single probability measure Pr. Assuming that ϕ and ψ are indeed included in Pr's domain, Pr assigns real values to $Pr(\phi)$ and $Pr(\psi)$. Again, however, this is not to say that the agent who has the salient comparative confidence ought to have confidences quantifiable using these precise values. Imprecise confidences, once formalized, express constraints on or properties of a probability distribution without determining the precise distribution in full detail.

An alternate approach to formalizing imprecise confidences that we will sometimes favor (e.g., in §2.2.1) instead uses *sets of probability distributions*. For example, on this approach, "I'm more certain of ϕ than ψ" is formalized as the *set* of distributions that satisfy this inequality: $\{Pr_i : Pr_i(\phi) > Pr_i(\psi)\}$. The difference between the two approaches can be clarified using more examples. The judgment "ϕ is more likely true than not" can be thought of as expressing a property of a single distribution ($Pr(\phi) > Pr(\neg\phi)$, or equivalently $Pr(\phi) > 1/2$) or it can be formalized as the set of distributions for which this property holds ($\{Pr_i : Pr_i(\phi) > 1/2\}$). The expressed confidence "ϕ and ψ are equally plausible" is formalized as the constraint $Pr(\phi) = Pr(\psi)$ on the first approach or, using the second approach, as $\{Pr_i : Pr_i(\phi) = Pr_i(\psi)\}$.

The approach we take in choosing how to formalize imprecise confidences makes a subtle difference in how we check or evaluate **Consistency**. If we follow the first approach, formalizing an agent's confidences as constraints on a single probability distribution, then an agent's confidences are consistent (i.e., **Consistency** is satisfied) when the constraints they convey are jointly satisfiable within a single probability distribution. If we instead formalize imprecise confidences as corresponding sets of probability distributions, then all of an agent's confidences together may be represented as the set-theoretic intersection of their formalizations, denoted $\mathcal{P}$.[17] In this case, an agent's confidences are consistent (i.e., **Consistency** is satisfied) when $\mathcal{P}$ is non-empty – that is, when there is at least one probability distribution that is a member of all of the sets that formalize that agent's respective confidences. In either case, demonstrating that confidences are *inconsistent* amounts to showing that their formalizations, at whatever level of precision, already imply a break with the axioms of probability theory.[18]

Consistency's notion of *consistency* is, of course, distinct from classical logic's. But unsurprisingly (since we've shown that probability generalizes on a certainty-functional reading of classical logic), the two are related. Extreme confidences (certainties of truth and falsity, formalized with values 1 and 0 respectively) are consistent [inconsistent] according to **Consistency** if and only if they are deductively consistent [inconsistent] – on the certainty-functional reading of classical logic.

1.4.3 Toward a Positive Defense

According to **Consistency**, confidences fit together logically (are consistent with one another) if they can be jointly represented within a single probability distribution. But why think that it's appropriate to turn to the mathematics

---

[17] van Fraassen (1984) calls $\mathcal{P}$ an agent's "representor" and Joyce (2010) calls it the agent's "credal committee."

[18] It's interesting to compare **Consistency** with Titelbaum's (2013, p. 81) stronger "Evaluative Rule" for "ranged attitudes," which would imply that imprecise confidences "violate the requirements of ideal rationality" just in case there is some precisification of any one of them that implies a break with the probability axioms. On this criterion, the imprecise confidences that $A \wedge B$ is more likely than not and A is at most 90% likely would be irrational (i.e., they would part from ideal rationality) since one could precisify the first of these using the assignment $Pr(A \wedge B) = .91[> .90 \geq Pr(A)]$ (it makes no difference that one could also precisify it using an assignment that doesn't imply a break with the probability axioms). For this reason, Titelbaum's Evaluative Rule would seem too strong at least to explicate the conditions under which imprecise confidences are consistent.

of probability when articulating the notion of consistency for confidences? Is there any compelling, positive case for believing that the logical coherence of confidences is appropriately described and legislated by the probability calculus?

We have already seen one reason for thinking this: probability theory is a conservative, straightforward generalization of classical, deductive logic for dealing with cases of uncertainty. To the extent that classical logic offers a satisfactory logic of certainty, we want it to be preserved in the limiting cases of a more general logic of confidence. Probability theory offers such a generalization by simply adapting deductive logic's bivalent semantics of certainty to gradational cases of uncertainty.

This argument, though informal, bears a strong resemblance to a more formally rigorous case appealing to Cox's well-known representation theorem (Cox 1946, 1961; cf., Jaynes 2003, ch. 2). Cox's controversial theorem (Halpern, 1999a,b; Colyvan, 2004, 2008) attempts to show that any real-valued measure of confidence built upon the foundation of classical, propositional logic ("the logical rules of Boolean algebra") will be isomorphic to a probability function so long as the following two requirements are met: (i) an agent's confidence in any negation $\neg\phi$ is a monotonically decreasing function of that agent's confidence in ϕ; and (ii) an agent's confidence in any conjunction $\phi \wedge \psi$ is a function of that agent's confidences in either conjunct (say ϕ) and in the other conjunct "with the additional assumption that" the first conjunct is true $(\psi|\phi)$.

Requirement (i) is a weakening of **Negation**, making the very plausible demand that greater confidence in any proposition corresponds with lesser confidence in that proposition's negation. Requirement (ii) also looks highly plausible, especially when comparing it to alternatives. For example, one's confidence in $\phi \wedge \psi$ does not follow from corresponding, individual confidences in ϕ and ψ. To adapt an example from Jaynes's (2003, §2.1) extended argument for (ii), I may be about equally confident in the three claims:

A: The next person I encounter will have blonde hair.

B: The next person I encounter will have a blue left eye.

C: The next person I encounter will have a brown right eye.

Nonetheless, my confidence in $A \wedge B$ is much greater than my confidence in $B \wedge C$, a difference that I take to be entirely consistent with my individual confidences in A, B, and C. This difference makes sense if confidences in conjunctions are rather determined as in requirement (ii) – since, for example, my confidence in A given B is much greater than my confidence in B given C.

The informal argument motivating our turn to probability and the formal case from Cox's theorem both build directly on the foundations of classical logic; they both assume that we want to preserve this logic in cases of certainty and to generalize from this logic for cases of non-extreme confidence. This is an important point as it serves to highlight the conditional nature of both cases for a probabilistic logic of confidence. Because they both assume and build upon classical logic, these arguments are compelling only at best for contexts in which classical logic is worth preserving. For "non-classical" contexts (plausibly including those involving vague propositions, fictional entities, self-referential paradoxes, quantum phenomena, etc.), we cannot safely assume the starting premise of these arguments – a point made very nicely by Colyvan (2004, 2008).

For additional arguments for **Consistency**, we may appeal to standard defenses of **Probabilism**, the related principle stating that an agent's confidences at a time ought to satisfy the axioms of probability.[19] The Dutch Book argument (Vineberg, 2016) constitutes perhaps the most popular defense of **Probabilism**. This argument helps to clarify a more precise sense in which we can see that non-probabilistic confidences (i.e., confidences whose formalizations are not jointly, consistently entailed by a single probability distribution) lack an internal coherence.[20] The Dutch Book argument reveals that non-probabilistic confidences lead us simultaneously to treat the same option differently when making choices. The argument explicitly deals with an agent's assessments of certain gambling scenarios. In the relevant scenarios, two bettors chip in possibly different amounts of money in order to make up the total stake; whoever wins the bet, takes the total stake – for a net payoff of the total stake minus what they originally contributed to the stake in order to enter the gamble.

As an example of the Dutch Book argument at work, say that an agent announces the following: "ϕ and $\neg\phi$ are both highly likely, neither having less than a 9 in 10 chance of being true." The expressed confidences can be formalized as follows: $Pr(\phi) > .9$ and $Pr(\neg\phi) > .9$. These confidences cannot be satisfied within the same probability distribution, of course, since they already imply a breaking of **Negation**. And thinking about how these confidences

---

[19] Though closely related, I prefer to distinguish the principle I've called **Consistency** from **Probabilism**. The most important difference between these is that **Probabilism** proposes a *necessary* condition on *rational* confidences whereas **Consistency** puts forward a *necessary and sufficient* condition for *logically consistent* confidences. Given **Consistency**, **Probabilism** might be restated as saying that rational confidences must be consistent.

[20] In fact, this is the rationale given explicitly by Ramsey (1926, p. 78) when he clarifies a sense in which "the laws of probability are laws of consistency, an extension to partial beliefs [confidences] of formal logic, the logic of consistency."

affect our assessments of options in gambling scenarios reveals a more specific sense in which they thereby run contrary to one another. Given $Pr(\phi) > .9$, the agent will consider a gamble advantageous that allows one to bet on ϕ for 75% of the total stake; but the agent is also highly confident in $\neg\phi$ ($Pr(\neg\phi) > .9$) and will thus consider a gamble advantageous that allows one to bet on $\neg\phi$ for 75% of the total stake. According to the agent's first confidence of $Pr(\phi) > .9$, it would be highly disadvantageous to have to contribute 75% of the total stake to bet on $\neg\phi$; but according to the agent's second confidence of $Pr(\neg\phi) > .9$, it would be highly disadvantageous to have to contribute 75% of the total stake to bet on ϕ. The agent's confidences are leading to contrary assessments of the same options.

As Ramsey puts the point, "If anyone's mental condition violated [the laws of probability], his choice would depend on the precise form in which the options were offered him, which would be absurd." The absurdity is highlighted by Ramsey by pointing out, in the next line, that such an agent "could have a book made against him by a cunning bettor and would then stand to lose in any event." If the agent in the previous paragraph were inclined to betting, for example, then he could be induced – based on his first confidence – to pay $75 for a chance at $100 if ϕ and – based on his second confidence – to pay $75 for a chance at $100 if $\neg\phi$. The cunning bettor would gladly allow him to make both bets, the agent paying $150 total for a guaranteed "winnings" of $100 whether ϕ or $\neg\phi$. The bettor takes $50 from the agent in any event.

While the Dutch Book argument reveals a specific sense in which non-probabilistic confidences lack coherence with one another, one might worry that this is a different sort of coherence than logical coherence. The notion of coherence at work arguably has to do with an agent's treatment of choices when making decisions; correspondingly, the negative repercussions of being incoherent in this sense has to do with ways in which our choices could lead to certain losses. If this is right, then the Dutch Book argument at best shows that non-probabilistic confidences may lead to imprudent actions, but it's not clear that this argument shows any sense in which such confidences are illogical (Christensen, 1996; Joyce, 1998).

This leads to another thought drawing upon the parallel case of deductive reasoning. In a deductive logic of consistency, the propositional beliefs of an agent at a time are deemed consistent if and only if the propositions believed could all simultaneously be true. Inconsistent beliefs are a problem at least because some of them have to be false; purely as a matter of logic, there is no chance that such beliefs could all jointly describe the world.[21]

---

[21] At least this is true in classical contexts. Priest (1979, 2002, 2006) has argued for *dialetheism*, the view that there are true contradictions at work, for example, in certain self-referential paradoxes.

Might we say something parallel in defense of **Consistency**'s account of logical fit between propositional confidences as joint explicability within a single probability distribution? If an agent's confidences don't satisfy this condition, is there a similar sense in which there's not a chance that they could jointly be oriented toward the world? It's easy to think of an intuitive sense in which this is the case, at least with certain non-probabilistic confidences. Say, for example, that an agent announces the following: "ϕ is more likely than not, and so is $\neg\phi$." The expressed confidences can be formalized as follows: $Pr(\phi) > .5$ and $Pr(\neg\phi) > .5$. Obviously, these confidences cannot be satisfied within any one probability distribution. And there does seem to be an obvious problem with this person's confidences, very much akin to the extreme case in which someone would announce certain belief in both ϕ and $\neg\phi$. Being more confident than not that both ϕ and $\neg\phi$ seems to betray an underlying misunderstanding that both ϕ and $\neg\phi$ could be simultaneously true. There's a sense in which such confidences cannot be simultaneously properly oriented toward the world. The world is such that either ϕ or $\neg\phi$; even if we have no evidence regarding which way the world is, we can already affirm that only at most one of the agent's two expressed confidences can be better aligned with the world.

As another example, an agent might judge that ϕ and ψ each have no more than a one in ten chance of being true, while also being more confident than not that $\phi \lor \psi$. The agent's confidences can be formalized as follows: $Pr(\phi) < .1$, $Pr(\psi) < .1$, and $Pr(\phi \lor \psi) > .5$ (implying a break with **General Additivity**). These confidences intuitively don't sit well together; there seems to be something wrong with these confidences such that they don't stand a chance of being jointly properly oriented toward the world. After all, the only way that $\phi \lor \psi$ could be true is if ϕ or ψ (or both) turn out true. So the agent's low confidences in these individual propositions don't sit well with a relatively high confidence in their disjunction. The world aligns well with lower confidences in ϕ and ψ only if it also aligns well with a low confidence in $\phi \lor \psi$. There is no possible way the world could be such that this agent's confidences could all collectively be sensibly oriented toward it.

That's the intuitive rationale. Is there a way of making it more rigorous? So-called epistemic utility arguments for **Probabilism** offer a promising resource here. Starting with the work of Joyce (1998, 2009) – himself building on yet more foundational work by de Finetti (2017) and Rosenkrantz (1981) – philosophers offer arguments showing that non-probabilistic confidences are necessarily dominated in terms of expected accuracy by probabilistically explicable confidences. Recent work by Pettigrew (2016, 2019) constitutes the state of the art here (see also Leitgeb and Pettigrew 2010a,b). Discussions of these arguments focus especially on the precise formal notions of accuracy and dominance at work. With these concepts explicated precisely, such arguments articulate exact senses in which non-probabilistic confidences cannot collectively align as neatly with the world (regardless of how the world actually is) as probabilistically formalizable confidences. Such arguments thus provide a strong reason to think that confidences that break with probability theory run into the same kind of trouble as beliefs that break with classical logic. And that supports **Consistency**, since it suggests that probability theory does indeed describe a notion of logical consistency for confidences – very much akin to classical logic's notion of consistency for beliefs.

This closing discussion has summarized some of the promising lines of argument for **Consistency**. Unsurprisingly, there is no accepted, airtight argument for this principle. And so, to some extent, we must seek out its proof in the pudding. Throughout the rest of this Element, we'll accept **Consistency** as a working principle, continually checking probability theory's plausibility and merits as a foundational logic of consistency for confidences.

2 Bayesian Epistemology

Section 1 introduced probability theory as a promising formal logic of consistency for confidences. This section now turns to an investigation of the epistemology of confidences: the study of what it takes for our confidences to be properly tuned to our experience of reality. Since, as we've argued, inconsistency (of beliefs or confidences) indicates a fundamental failure of one's epistemic attitudes to be collectively oriented to reality, the logic of consistency is relevant to epistemology. It's nonetheless important to remember the distinction between the two fields.

The logic of consistency guides us in forming epistemic attitudes that fit together well; however, strictly speaking, it's silent when it comes to leading us to have beliefs, confidences, and so on that are properly informed by, and sensitive to, our experience of the world outside of our heads. Nobody perhaps makes this point so delightfully as G. K. Chesterton. In a chapter entitled "The Maniac," Chesterton (1908, p. 32) effectively argues that nobody is more logical than a madman:

> [The madman's] most sinister quality is a horrible clarity of detail; a connecting of one thing with another in a map more elaborate than a maze. If you argue with a madman, it is extremely probable that you will get the worst of it; for in many ways his mind moves all the quicker for not being delayed by the things that go with good judgment. He is not hampered by a sense of humour or by charity, or by the dumb certainties of experience. He is the more logical for losing certain sane affections. Indeed, the common phrase for insanity is in this respect a misleading one. The madman is not the man who has lost his reason. The madman is the man who has lost everything except his reason.

Clearly, we need to go beyond a logic of consistency ("a connecting of one thing with another in a map more elaborate than a maze") if we want to sort out what it takes for beliefs, confidences, and the like to be reasonable, not just with respect to each other, but also with respect to our experience of the world. This same point is made, albeit in a less entertaining way, by the two acknowledged forefathers of modern Bayesianism, Bruno de Finetti[22] and Frank Ramsey.[23]

It's time we focused on epistemology. To borrow Chesterton's phrase, this section is concerned with the ways in which our confidences should be hampered by the dumb certainties of experience. As we proceed, we'll find that the Bayesian interpretation of probability doesn't just provide a promising logic of confidence, but also a fruitful formalism for developing an epistemology of confidence. While the probability calculus – properly interpreted – is itself supposed to provide the logic of consistency for confidences, the epistemology of confidences uses the probability theory as a tool for articulating epistemological principles not already encompassed by probability theory. Let's begin by motivating the project with a glimpse into the work of Bayesian epistemology's patron saint.

2.1 Bayes and Bayesian Epistemology

The Reverend Thomas Bayes's seminal "Essay Towards Solving a Problem in the Doctrine of Chances" (1763) was published two years after his death. Bayes's past acquaintance and foremost supporter, Richard Price, presented the Essay (complete with Price's own lengthy commentary) to the Royal Society. In the Essay, Bayes sets out to solve the following problem:

> *Given* the number of times in which an unknown event has happened and failed: *Required* the chance that the probability of its happening in a single trial lies somewhere between any two degrees of probability that can be named.

Two considerations in particular highlight the epistemological nature of Bayes's problem. First, while questions of probability among Bayes's peers

---

22 *"The calculus of probability can say absolutely nothing about reality* [...] One can express in terms of it any opinion whatsoever, no matter how "reasonable" or otherwise, and the consequences will be reasonable, or not, for me, for You [sic], or anyone, according to the reasonableness of the original opinions of the individual using the calculus. As with the logic of certainty, the logic of the probable adds nothing of its own: it merely helps one to see the implications contained in what has gone before" (de Finetti, 2017, p. 182; emphasis in original).

23 "[W]e do not regard it as belonging to formal logic to say what should be a man's expectation of drawing a white or a black ball from an urn; his original expectations may within the limits of consistency be any he likes; all we have to point out is that if he has certain expectations he is bound in consistency to have certain others. This is simply bringing probability into line with ordinary formal logic, which does not criticize premises but merely declares that certain conclusions are the only ones consistent with them" (Ramsey, 1926, p. 85).

and predecessors largely concerned the implications of known probabilities on future experiments and observations, Bayes's is the inverse problem: given the results (specified numbers of successes and failures) of a number of trials of an experiment, he seeks the probabilities (within certain constraints) of the possible outcomes. Second, Bayes's working definition of probability is also unmistakably epistemic, associating probabilities with confidences or "expectations."[24]

With this interpretation of probability in mind, Bayes's problem falls squarely into the realm of epistemology; he is inquiring as to how our experiences of the world ought to constrain and shape our expectations or confidences about the world. As Price (Bayes, 1763, p. 406) puts it:

> Every one sees in general that there is reason to expect an event with more or less confidence according to the greater or less number of times in which, under given circumstances, it has happened [...]; but we here see exactly what this reason is, on what principles it is founded, and how we ought to regulate our expectations.

Bayes's solution draws upon at least three important epistemological postulates. The first describes how an agent's confidences should change over time as he or she learns new information. The second expresses how agents' rational confidences ought to be calibrated in line with particular modes of evidence. The third postulate is a constraint on how confidences at a time should relate to one another, when these pertain to any "unknown event," or "an event concerning the probability of which we absolutely know nothing antecedently to any trials made concerning it" (pp. 392–393). These postulates and the developments in Bayesian epistemology that they've since inspired will structure our discussion throughout the rest of this section.

2.2 Bayes's Rule

In order to understand Bayes's first epistemological postulate, it's instructive to compare two of the "Propositions" that he offers near the beginning of the

---

[24] Bayes writes, "The *probability of any event* is the ratio between the value at which an expectation depending on the happening of the event ought to be computed, and the value of the thing expected upon it's [sic] happening." Price makes the following remark concerning this definition: "Instead therefore, of the proper sense of the word *probability*, [Bayes] has given that which all will allow to be its proper measure in every case where the word is used." Exactly what Price considered to be the proper sense of the word is left unclear, but if he's right that Bayes was offering a measure here as opposed to a strict definition, then the idea resonates with modern epistemic interpretations of the sort we've put forward in Section 1. The ratio of the amount at which an event's expectation ought to be valued to the value gained if the event occurs sounds a lot like "fair betting quotient" measure of confidence (Earman, 1992, p. 8). This ratio measures (more or less accurately) the strength of that expectation itself just as betting quotients may provide "a measure of the confidence of my opinion" (Ramsey, 1926, pp. 70–72).

Essay.[25] In Proposition 3, Bayes introduces and attempts to demonstrate **Conditional Probability**; for two subsequent events, the probability the second occurs "on supposition the 1st happens" is $Pr(E_2|E_1) = Pr(E_1 \wedge E_2)/Pr(E_1)$. By contrast, Bayes's Proposition 5 puts forward the following as an additional, distinct posit:

> If there be two subsequent events, the probability of the 2d $[Pr(E_2)]$ and the probability of both together $[Pr(E_1 \wedge E_2)]$, and it being 1st discovered that the 2d event has happened, from hence I guess that the 1st event has also happened, the probability I am in the right is $[Pr(E_1|E_2) = Pr(E_1 \wedge E_2)/Pr(E_2)]$.

Proposition 5 may at first look redundant, simply spelling out another instance of **Conditional Probability**. But viewing it in this way ignores an important subtlety in Bayes's presentation. Bayes's Proposition 3 derives a probability on the "supposition" that an event occurs, whereas Proposition 5 pertains to the case where an agent "discovers" or learns that an event occurs. The fact that Bayes resolves these propositions in mathematically equivalent ways, as applications of **Conditional Probability**, shouldn't blind us to the fact that the probabilities in question are conceptually distinct. The probability of some ϕ on the supposition that ψ, $Pr(\phi|\psi)$, is distinguishable from the probability of ϕ *having learned that ψ*. While the former's derivation is provided by the rules of probability theory (interpreted epistemically), the latter is a matter of epistemology. How should confidence in some ϕ be affected upon learning that ψ is true? Bayes's Proposition 5 provides an answer to this epistemological question, asserting that ϕ's probability upon discovering that ψ should be determined by a suppositional (conditional) probability $Pr(\phi|\psi)$ – of the sort introduced in Proposition 3.

The answer given by Bayes, now known as **Bayes's Rule**, may be stated more precisely as follows:[26]

Bayes's Rule. Upon learning ψ (and nothing else), the new probability of any proposition ϕ should equal the old (just previous to learning ψ) conditional probability of ϕ given ψ: $Pr_{new}(\phi) = Pr_{old}(\phi|\psi)$.

It is crucial to distinguish **Bayes's Rule**, the epistemological postulate, from **Bayes's Theorem**, the theorem of probability we introduced in §1.3.3.

---

[25] Throughout this discussion, I have replaced Bayes's formal notation with our own formal explication of his remarks.

[26] Depending on how conditional probabilities are defined (see Section 1, footnote 13), **Bayes's Rule** may only pertain to situations in which we've learned evidence that had some positive probability; i.e., in the formal statement above, for which $Pr_{old}(\psi) > 0$. **Bayes's Rule** is also known in the literature as "Strict Conditionalization," "Bayesian Conditionalization," or often just "Conditionalization."

To see **Bayes's Rule** in action, imagine that before learning new information you were equally confident that any of the cards found in a standard 52-card deck has been drawn – knowing with certainty that exactly one such card has been drawn from the deck. This past confidence that the Jack of Hearts has been drawn in particular $[H_J]$ is formalized as $Pr_{old}(H_J) = 1/52$. Moreover, on the supposition that the card was drawn by your card sharp friend Lucky $[L]$, you were substantially more confident that it's the Jack of Hearts: $Pr_{old}(H_J|L) = 1/12$. Given your past confidences, what should your present confidence in H_J be in light of your recent discovery that Lucky did indeed draw the card? The answer need not be determined by $Pr_{old}(H_J|L) = 1/12$. While your suppositional confidence prior to the discovery was $1/12$ in this proposition, one might think it epistemically permissible for *supposing* to have an influence on one's confidences distinct from that of *learning*.[27] **Bayes's Rule**, however, requires these influences to be identical:[28] $Pr_{new}(H_J) = Pr_{old}(H_J|L) = 1/12$.

2.2.1 Interlude: Updating Imprecise Confidences

How might **Bayes's Rule** be adapted for scenarios in which agents seek to update quantifiably *imprecise* confidences (recall §1.4)? The rule's underlying core idea that updated confidences should be determined by prior, suppositional confidences pertains equally well to precise and imprecise confidences. An agent might be equally confident that either one of two candidates will win the election, while having a higher though quantitatively imprecise confidence that candidate A will win supposing that A wins the state-wide vote in Arizona. A generalized version of **Bayes's Rule** could then require that this agent, upon learning that A does indeed win in Arizona, update to his or her prior, imprecise, suppositional confidence that candidate A is more likely to win.

A more exact statement of such a generalization of **Bayes's Rule** is attainable if we formalize an agent's confidences using a set of probability functions $\mathcal{P}$ (as in §1.4.2). From this perspective, the updating of imprecise confidences can be thought of as applying **Bayes's Rule** individually to each member of a set

---

27 Cf. Ramsey's brief remark (1926, p. 76): "[T]he degree of [an agent's] belief in p given q [...] is not the same as the degree to which he would believe p, if he believed q for certain; for knowledge of q might for psychological reasons profoundly alter his whole system of beliefs."

28 It's easy to prove that **Bayes's Rule** must be true so long as $Pr_{new}(\psi) = 1$ and $Pr_{new}(\phi|\psi) = Pr_{old}(\phi|\psi)$. Upon learning some ψ with certainty then, if one's quantifiably precise confidences are updated in a way that breaks with **Bayes's Rule**, this implies that $Pr_{new}(\phi|\psi) \neq Pr_{old}(\phi|\psi)$. As Jonathan Livengood has pointed out to me, this could be taken to provide formal backing and precision for Ramsey's notion of a "whole system of beliefs" being "profoundly altered" (see the previous footnote). On this interpretive question in Ramsey, also see (Levi, 2004, p. 468).

of probability functions. Accordingly, updated confidences are identified with the appropriate set of conditional probabilities:

Bayes's Rule (for sets of probabilities). Upon learning ψ (and nothing else), an agent's new confidence in any proposition ϕ, $\mathcal{P}_{new}(\phi)$, should equal his or her old (just previous to learning ψ) confidence in ϕ conditional on ψ, $\mathcal{P}_{old}(\phi|\psi)$, that is, the set of old conditional probabilities of ϕ given ψ: $\mathcal{P}_{new}(\phi) = \{Pr_{old}(\phi|\psi) : Pr_{old} \in \mathcal{P}_{old}\} = \mathcal{P}_{old}(\phi|\psi)$.

In the preceding example, the agent is initially equally confident that either candidate will win the election. Every probability function that is a member of this agent's $\mathcal{P}_{old}$ thus agrees that $Pr_{old}(A) = Pr_{old}(B)$. The agent might additionally have the initial confidence that there is at least a 9 in 10 chance that either candidate A or B will win the election (no third-party candidate will win instead). Then, given **Additivity**, every probability function that is a member of $\mathcal{P}_{old}$ will agree that $Pr_{old}(A) = Pr_{old}(B) \in [.45, .5]$. Or perhaps the agent is instead additionally certain that one and only one of these two candidates will win the election, in which case all members of $\mathcal{P}_{old}$ will agree that $Pr_{old}(A) = Pr_{old}(B) = .5$. It is consistent with all of this that, on the supposition that candidate A wins Arizona (Z), the agent is more confident that A will win without any precise sense of how much more confident; in this case, all members of $\mathcal{P}_{old}$ agree that $Pr_{old}(A|Z) > Pr_{old}(B|Z)$ and thus that $Pr_{old}(A|Z) \in (.5, 1]$.

This example shows that by following **Bayes's Rule (for sets of probabilities)**, an agent's confidences can actually become less precise upon learning new evidence. Thus, the agent in our present example might have imprecise confidence $\mathcal{P}_{old}(A) = [.45, .5]$ or even quantifiably precise confidence $\mathcal{P}_{old}(A) = \{.5\}$, only to lose a great amount of precision by updating on new evidence $\mathcal{P}_{new}(A) = \mathcal{P}_{old}(A|Z) = (.5, 1]$. It turns out that this can occur, in some cases (plausibly including the present example), upon learning *any* member of a partition of possible evidence – so that, for example, one's confidences can become less precise upon performing a successful experiment, no matter which of the possible results of the experiment is observed (Seidenfeld and Wasserman, 1993). Such "dilation" is sometimes thought to pose a serious problem for **Bayes's Rule (for sets of probabilities)** or for the general set-based approach to explicating imprecise confidences; prima facie at least, it seems counterintuitive that a strict increase in an agent's evidence results in his or her confidences being *less* precise. However, it is a matter of ongoing debate whether dilation truly poses a serious problem for this account of imprecise confidences (Seidenfeld and Wasserman, 1993; Bradley and Steele, 2014; Pedersen and Wheeler, 2014).

2.2.2 Complications and Generalizations

Before discussing the general epistemological merits of **Bayes's Rule**, it's worth highlighting a couple of awkward consequences that follow from this postulate and how the rule has been adapted accordingly. First, **Bayes's Rule** presumes that learning some proposition ψ amounts to placing absolute certainty in ψ. After all, one of the propositions an agent must update his or her confidence in, in light of learning some ψ, is ψ itself. Applying **Bayes's Rule** to this proposition: $Pr_{new}(\psi) = Pr_{old}(\psi|\psi) = 1$. But while it's perhaps acceptable to think that we do sometimes learn new information with certainty, it seems manifestly true that we sometimes have less than full certainty in the propositions we've learned. For example, Ramsey (1926, p. 86) mentions cases in which "I think I perceive or remember something but am not sure." Richard Jeffrey (1983a, p. 135) argues that such cases are "typical of our most familiar sorts of updating, as when we recognize friends' faces or voices or handwritings pretty surely, and when we recognize familiar foods pretty surely by their look, smell, taste, feel, and heft." What do we do, in the previous example, when we don't learn with certainty that our friend Lucky drew the card, but we only learn this pretty surely – perhaps because we only caught a quick glimpse of his face?

Bayes's Rule is apparently unable to guide such common revisions to our confidences. When an experience shifts our confidence in some ψ, but not all the way up to certainty, how should that shift in confidence influence our other confidences? Jeffrey (1983b) develops the following generalization of **Bayes's Rule**[29] as an answer:

Jeffrey's Rule. When a learning experience's full, direct effect amounts to shifting one's confidences in the members of a partition $\{\psi_i\}$ from $Pr_{old}(\psi_i)$ to $Pr_{new}(\psi_i)$, the new probability of any proposition ϕ should be the following weighted average of the old conditional probabilities of ϕ given each ψ_i: $Pr_{new}(\phi) = \sum_i Pr_{old}(\phi|\psi_i)Pr_{new}(\psi_i)$.

Developing the previous example, suppose again that my initial confidence the Jack of Hearts is drawn is $Pr_{old}(H_J) = 1/52 \approx .02$. Moreover, say my initial confidence that Lucky draws the card is $Pr_{old}(L) = 1/100$. This

---

[29] Note that, in the special case when there is only one member of the partition learned with certainty, $Pr_{new}(\psi) = 1$, **Jeffrey's Rule** is equivalent to **Bayes's Rule**.

confidence increases dramatically, when I catch a quick glimpse of the person drawing the card, up to $Pr_{new}(L) = 8/10$. Given that $Pr_{old}(H_J|L) = 1/12[= Pr_{new}(H_J|L)]$, I ought to be more confident now that the Jack of Hearts is drawn. How much more confident? Using $\{L, \neg L\}$ as my "evidence partition," **Jeffrey's Rule** entails $Pr_{new}(H_J) = Pr_{old}(H_J|L)Pr_{new}(L) + Pr_{old}(H_J|\neg L)Pr_{new}(\neg L) \approx .07$.

Jeffrey's Rule pertains only to cases in which all probabilities conditional on members of the evidence partition are *rigid* across the update; that is, for any proposition ϕ, $Pr_{old}(\phi|\psi_i) = Pr_{new}(\phi|\psi_i)$ across all members of the partition $\{\psi_i\}$. In fact, given this rigidity condition, **Jeffrey's Rule** follows directly from the probability calculus. However, as Pearl (1988, §2.3.3) and McGrew (2010, 2014), among others, have pointed out, rigidity is an extremely difficult and complicated condition to assess. In checking for rigidity, it is tempting to rely purely on one's intuitions. But cases in which the salient conditional probabilities intuitively appear to be rigid can often be shown to break with this condition when they're thought through more carefully and in the light of more details.

As a case in point, are the relevant probabilities truly rigid in the example above? Plausibly, $Pr_{new}(H_J|L) = Pr_{old}(H_J|L)$ and $Pr_{new}(H_J|\neg L) = Pr_{old}(H_J|\neg L)$. However, if in addition to raising my confidence that Lucky is the dealer, my quick glance at the dealer also boosts my confidence that the dealer is Ace [A] (who looks a bit like Lucky in the relevant respects), then rigidity plausibly fails since the "passage of experience" has made it much more likely that the dealer is Ace conditional on the dealer not being Lucky: $Pr_{new}(A|\neg L) \gg Pr_{old}(A|\neg L)$. Moreover, if Ace is also a card sharp (with an inclination for or against the Jack of Hearts), then this failure of rigidity apropos A and $\neg L$ may directly call into question the previously intuitive rigidity apropos H_J and the evidence partition: $Pr_{new}(H_J|\neg L) \neq Pr_{old}(H_J|\neg L)$.

These points were not lost on Jeffrey, who specifies that the evidence partition should include *all* propositions whose probabilities were directly affected by the experience. Now that we know about Ace, we would want to update using the finer-grained partition $\{L, A, \neg(L \vee A)\}$. Nonetheless, the above example speaks to the volatility of the rigidity condition. Passages of experience directly affect our confidences in a multitude of propositions. For rigidity to be satisfied, our evidence partition must explicitly span those propositions. To have any hope of accurately uncovering such a partition, it would certainly help to know exactly what it takes for a passage of experience to influence a proposition's confidence *directly*. However, the practical problem seems to run even deeper than this, since Pearl (1988, pp. 66–67) shows that rigidity can still fail, even with respect to propositions that are plausibly *not* directly affected by the passage of experience. The search for an evidence partition that has a hope of securing rigidity thus goes beyond the already daunting task of articulating all propositions directly affected by an experience. The upshot is that **Jeffrey's Rule**, while on secure mathematical footing, seems far more limited in its scope and applicability than philosophers have thought and hoped.

Another unappealing consequence of **Bayes's Rule** is that certainties cannot become uncertainties, on pain of irrationality. Extreme confidences (1s and 0s) are, in this sense, "sticky," since no amount of updating using **Bayes's Rule** (or **Jeffrey's Rule**) can convert extreme probabilities to non-extreme.[30] By contrast, it seems that agents may be certain about propositions at one point in time and then lose that certainty at a later point in time without thereby becoming irrational. Perhaps the most obvious instances motivating this thought involve simple forgetting. We're constantly forgetting things that we once took for certainties. Right now, as I write this, I'm certain that I'm typing out this sentence. But I don't plan on remembering that this was occurring at that indexed point in time for the rest of my life. Neither do I think this is a mark of irrationality; humans aren't irrational because they forget things.

There are two common responses to this challenge on behalf of Bayesianism. First, one might leave **Bayes's Rule** as is, and focus on offering principled restrictions on how agents may rationally assign extreme probabilities. If extreme probabilities mark an unalterably dogmatic certainty with respect to the truth or falsity of a proposition, then perhaps we only need to specify the class of propositions for which rationality allows and requires such dogmatism. In fact, the **Normality** axiom (§1.2) already does some of this work for us, specifying a class of propositions (i.e., the tautologies) with probability 1.[31]

---

30 If a proposition P has extreme probability $Pr(P) = 1$, then for *any* proposition ψ with non-extreme probability, **Total Probability** implies $Pr(P|\psi)Pr(\psi) + Pr(P|\neg\psi)Pr(\neg\psi) = 1$. Since the left side is a weighted average of $Pr(P|\psi)$ and $Pr(P|\neg\psi)$, this implies $Pr(P|\psi) = Pr(P|\neg\psi) = 1$.

31 Doesn't **Normality** then require actual agents to be logically omniscient, having full certainty in all logical truths? **Normality** makes no mention of actual agents at all; thus, if it does carry such implications, it is only in combination with a bridge principle specifying probability theory's relevance to the attitudes of actual agents. **Consistency** (§1.4.2) plays that role for us. But according to **Consistency**, an agent's confidence in any tautology **T** may be acceptable so long as it is entailed by $Pr(\mathbf{T}) = 1$. Apart from being certain about a particular tautology, there are at least two other ways an agent might satisfy this requirement: (1) He or she might not have any confidence whatever in the relevant tautology, perhaps because the agent has simply never pondered the tautology. (2) The agent's confidence about **T** is not quantitatively precise but is entailed by $Pr(\mathbf{T}) = 1$. For example, being unsure as to whether tautology **T** is logically true, logically false, or contingent, an agent's confidence in **T** may be no more precise than $0 \leq Pr(\mathbf{T}) \leq 1$. Or thinking that the proposition is likely true and indeed could be a tautology, the agent might have confidence $Pr(\mathbf{T}) > .5$. These formalizations are entailed by all probability distributions (i.e., by $Pr(\mathbf{T}) = 1$) and so are permitted by **Consistency**. That said,

Normality and **Negation** moreover imply that logical falsehoods (negations of tautologies) must take probability 0. But while these rules inform us that we should be dogmatic in our certainties about certain propositions, they don't specify the class of propositions about which we *shouldn't* be dogmatic.

The principle commonly used to do this is **Regularity**, which states that 1s [0s] are to be reserved for tautologies [logical falsehoods] exclusively. Stated as a principle of epistemic rationality: agents ought to be nondogmatic (less than fully certain, and so open to evidence-based revision of their confidences) in all of their confidences apart from the certainties they have in tautologies.

This principle is often called into question, as it certainly seems like there are logically contingent propositions that agents may rationally be certain about. I'm quite certain that no bachelors are married. I'm also certain that I'm not my mother. This worry may be alleviated, at least to some extent, by widening the scope of propositions referenced in **Regularity**. More lenient versions of **Regularity** may, for example, add analytic truths and metaphysically necessary truths into the realm of propositions that properly take probability one. These two addenda together would arguably bar the above counterexamples. And this move is well-motivated to the extent that one sees all of the propositions included in these categories as commonly deserving a uniquely dogmatic attitude of certainty, that is, to the extent that one agrees that these are the propositions about which it would indeed be irrational to leave open the door to less than extreme confidence.

But even granting that a widening of **Regularity** may more suitably capture a class of propositions that exclusively should be assigned extreme confidence, this principle runs into other trouble. For one thing, we were introducing **Regularity** as a means of making **Bayes's Rule** more palatable, when in fact the former principle severely restricts the application and importance of the latter rule. This is because, as we have just discussed, **Bayes's Rule** identifies learning new evidence with assigning probability one to that evidence. But then **Regularity** requires that we only use this rule for cases where we "learn" tautologies (or possibly other necessary truths). Moreover, **Normality** requires that tautologies be assigned probability one from the start. Since $Pr(\phi|\psi) = Pr(\phi)$ whenever $Pr(\psi) = 1$, Bayes's Rule judges these to be

---

the challenge remains, since some confidences that could seem permissible would inevitably lead to inconsistency; e.g., an agent's being equally confident between T and $\neg T$, $Pr(T) = .5$. The challenge of logical omniscience has been approached in a variety of ways in the literature, including approaches that break with **Normality** (Good, 1968; Garber, 1983; Titelbaum, 2013). Fully meeting the challenge may require combining the above considerations with such existing approaches.

degenerate cases in which no updating of confidences is allowed. The upshot is that **Regularity** renders **Bayes's Rule** useless and uninteresting – though one might hold out hope that **Jeffrey's Rule** can legislate updates in all interesting cases.

Even setting this concern aside, **Regularity** fails to offer a total solution to the problem in question. If this principle is taken to offer a total fix, then the problem in my quick example above is that I placed extreme confidence in a proposition that was not a logical truth (or analytic truth, or metaphysical necessity, etc.). It was irrationally dogmatic of me to be absolutely certain that I was writing that sentence at the time of my writing it. Since my confidence was not extreme, it was reducible by further updates following **Bayes's Rule** to an appropriate value corresponding to my later uncertainties concerning what exactly I was doing at that time on that day. However, **Bayes's Rule** has us update our confidences upon learning new information. Cases of rational forgetting, by contrast, are not cases in which confidences are affected by the intake of new information, but rather by the losing of old information.

This consideration in particular motivates a second possible response to the challenge of sticky certainties. Instead of trying to demarcate the class of propositions about which agents rationally may be certain, this response generalizes **Bayes's Rule** so that certainties are no longer sticky. Agents following such a generalized rule may be modeled as gaining and losing certainties over time. Inspired by and building upon similar ideas put forward by Levi (1980, 1987), Titelbaum (2013) develops and defends such a rule:

Titelbaum's Rule. The probabilities across times t_j and t_k of any proposition ϕ should be related as follows (assuming we are conditionalizing on propositions with positive probability): $Pr_j(\phi|\langle C_k - C_j \rangle) = Pr_k(\phi|\langle C_j - C_k \rangle)$.

C_i denotes the set of propositions the agent places full certainty in at time t_i. $C_i - C_l$ is the standard set-theoretic subtraction: the set containing all members of C_i that are not members of C_l. And the brackets $\langle \cdot \rangle$ denote a function from a set of propositions to a proposition equivalent to the conjunction of those propositions (where $\langle \emptyset \rangle$ returns a tautology). The idea here then is as follows: allow that an agent's certainties differ at times t_j and t_k; between these times, certainties may have been gained or lost (or both) by the agent. Then the agent's confidence in any proposition at t_j conditional on those certainties had at t_k (but not at t_j) should equal the agent's confidence in that proposition at t_k conditional on those certainties had at t_j (but not at t_k). **Titelbaum's Rule** thus requires a diachronic coherence between an agent's suppositional confidences at different times.

Since $\langle\emptyset\rangle$ is a tautology, we get the following special instance of **Titel-baum's Rule**: if $C_j \subseteq C_k$, $Pr_j(\phi|\langle C_k - C_j\rangle) = Pr_k(\phi)$. Notice that there is no specification about which time t_j or t_k comes earlier than the other. If t_j is the "old" time and t_k the "new," then this special case of **Titelbaum's Rule** is equivalent to **Bayes's Rule**. If, however, t_k is earlier than t_j, then this special case of **Titelbaum's Rule** speaks to a case in which the agent has only lost prior certainties. In that case, this rule mandates that the agent's new confidence *conditional on the certainties lost* should equal his or her old confidence (just prior to losing these certainties).[32]

2.2.3 Epistemology and Convergence

Bayes's Rule connects the confidences an agent ought to have after learning some proposition with the suppositional confidences that the agent had prior to this learning. As such, this rule provides dynamical constraints on an agent's rational confidences over time; given **Bayes's Rule**, agents can't just shift their confidences willy-nilly, but instead must fit any updated confidences to their prior, suppositional confidences. **Bayes's Rule** might plausibly be thought of as enforcing a diachronic version of coherence then. But how does **Bayes's Rule** fare as an epistemological principle? Does it appropriately legislate how our confidences ought to be shaped by our experience of the world?

In order to argue that it does, Bayesians appeal to well-known convergence or "swamping" theorems (Savage, 1954; Blackwell and Dubins, 1962; Gaifman and Snir, 1982). These theorems aim to demonstrate the objectifying influence an agent's evidence may have on his or her confidences over time. By updating confidences in accordance with **Bayes's Rule**, an agent's prior confidences – no matter how subjectively tainted initially – come to accurately represent objective reality. The initial subjectivity of such confidences is "swamped," as it were, by the agent's objective evidence. These theorems aim to show this by proving a related point about converging confidences: no matter the extent to which subjects disagree in their initial confidences, so long as they all follow **Bayes's Rule**, a sufficient amount of common evidence will inevitably lead to a convergence in their opinions (i.e., updated confidences).

A simple example serves to illustrate such convergence. There are two decks of cards: deck 1 is standard, deck 2 is heartless – that is, only including the 39 spades, clubs, and diamonds. One of these decks is discretely chosen, and cards are drawn with replacement from the chosen deck. Before any cards are

---

[32] For yet another example of how **Bayes's Rule** might be usefully generalized, in this case to accommodate explanatory judgments, the reader is referred to the recent work of Igor Douven (2013, 2020).

drawn (at time t_0), Dmitri is equally confident that each deck has been chosen: $Pr_{D,t_0}(H_1) = Pr_{D,t_0}(H_2) = .5$. By contrast, Xenia is nearly certain that the biased deck has been chosen: $Pr_{X,t_0}(H_1) = .001$, $Pr_{X,t_0}(H_2) = .999$. The first draw is a diamond $[D_1]$, and the agents follow **Bayes's Rule** in order to update their confidences accordingly:

$$Pr_{D,t_1}(H_1) = Pr_{D,t_0}(H_1|D_1)$$

$$= \frac{Pr_{D,t_0}(H_1)Pr_{D,t_0}(D_1|H_1)}{Pr_{D,t_0}(H_1)Pr_{D,t_0}(D_1|H_1) + Pr_{D,t_0}(H_2)Pr_{D,t_0}(D_1|H_2)}$$

$$= \frac{.5 \times 1/4}{.5 \times 1/4 + .5 \times 1/3} = .4286,$$

$$Pr_{X,t_1}(H_1) = Pr_{X,t_0}(H_1|D_1)$$

$$= \frac{Pr_{X,t_0}(H_1)Pr_{X,t_0}(D_1|H_1)}{Pr_{X,t_0}(H_1)Pr_{X,t_0}(D_1|H_1) + Pr_{X,t_0}(H_2)Pr_{X,t_0}(D_1|H_2)}$$

$$= \frac{.001 \times 1/4}{.001 \times 1/4 + .999 \times 1/3} = .0008.$$

The second draw is a spade $[S_2]$, leading to the next round of updates:

$$Pr_{D,t_2}(H_1) = Pr_{D,t_1}(H_1|S_2)[= Pr_{D,t_0}(H_1|D_1 \wedge S_2)]$$

$$= \frac{.4286 \times 1/4}{.4286 \times 1/4 + .5714 \times 1/3} = .36,$$

$$Pr_{X,t_2}(H_1) = Pr_{X,t_1}(H_1|S_2) = .0006.$$

After eight more draws, still not a single heart has been drawn:

$$Pr_{D,t_{10}}(H_1) = Pr_{D,t_2}(H_1|D_3 \wedge D_4 \wedge S_5 \wedge C_6 \wedge C_7 \wedge S_8 \wedge D_9 \wedge C_{10})$$

$$= \frac{.36 \times (1/4)^8}{.36 \times (1/4)^8 + .64 \times (1/3)^8} = .0533,$$

$$Pr_{X,t_{10}}(H_1) = .0001.$$

As their common evidence accumulates, the agents continue to converge in their respective confidences. For example, after ten more such draws, Dmitri and Xenia both should have confidence less than .003 that they are using the standard deck. As the number of heartless draws increases without bound, Dmitri and Xenia converge to full agreement, both approaching complete certainty that they are working with the biased deck. Of course, this convergence happens immediately if a single heart is drawn, in which case both agents instantly attain to certainty that they are working with the standard deck.

Convergence theorems invoke three conditions describing the type of agents needed for the result to hold (Earman 1992, p. 142; Sprenger and Hartmann 2019, p. 291):

1. An agent's confidences at any one time are represented by a single probability distribution; they are quantitatively precise, satisfying the axioms of probability.
2. All agents update their confidences (probabilities) in accordance with **Bayes's Rule**.
3. The agents have identical certainties; formally, they assign 1s and 0s to exactly the same propositions.

Regardless of how radically the initial (t_0) confidences of such agents diverge, as they are confronted with an ever-increasing line of common evidence and update these confidences, they are guaranteed in the long run to converge in their opinions.

While convergence theorems provide prima facie evidence for the epistemological significance of **Bayes's Rule**, they may be turned on their heads in order to show that this rule is not ultimately doing the epistemological heavy lifting. First, note that convergence is only guaranteed in the long run. When the long run is replaced by any finite string of common evidence, the results may ironically be restated as *non-convergence* theorems. Allow that conditions 1–3 above attain for two agents. Now instead of specifying "how radically the initial (t_0) confidences of these agents diverge," specify the finite extent of the common evidence they can mutually take in (this could, for example, correspond to the maximum amount of evidence an agent could glean in a lifetime, or even to the amount of evidence that could be collected and passed down throughout the history of the human race). Furthermore, stipulate some bounds of agreement δ within which we would allow that the agents agree in their opinions. Then we can mathematically set a range of initial priors for these two agents diverging to such a degree that the agents' opinions will not converge to within δ after taking in that amount of common evidence, despite the fact that the agents each satisfy 1–3. The upshot is that the convergence theorems don't guarantee convergence for finite agents in the actual world. In order to secure such convergence, at the very least, further constraints would have to be in place precluding agents from having arbitrarily divergent, initial opinions.[33]

---

[33] More sophisticated non-convergence results in the literature highlight conditions under which convergence may not be guaranteed or even possible in the long run (Feldman, 1991; Belot, 2013).

Table 1 Diverging confidences,
sequentially updated using **Bayes's Rule** and
given a common line of evidence.

	$Pr_{D,t_i}(H_1)$	$Pr_{X,t_i}(H_1)$
t_0	.5000	.0010
t_1	.6000	.0008
t_2	.6923	.0006
t_3	.7714	.0004
t_4	.8351	.0003
t_5	.8836	.0002
t_6	.9193	.0002
t_7	.9447	.0001
t_8	.9624	.0001
t_9	.9746	.0001
t_{10}	.9830	.0001

There is another serious concern with using the convergence theorems to defend the epistemological significance of **Bayes's Rule**. Far from demonstrating that this rule provides the desired fit between agents' confidences and the world as they experience it, convergence theorems presume such fit already exists. Recall that **Bayes's Rule** only requires updated confidences to fit with past, suppositional confidences. But these suppositional confidences may in principle be as subjectively tainted as any other. If these don't already encode a rational response to the supposed evidence, then there is no reason to think that credences updated on their basis will increasingly fit our experience of the world.

Without presuming that agents' suppositional confidences fit their evidence appropriately, they may satisfy conditions 1–3 and nonetheless even *diverge* in their opinions when updating on common evidence. In the above example, we assumed that the agents' confidences of drawing a particular card, supposing the deck to be standard or heartless, equaled the proportion of such cards in such a deck. But **Bayes's Rule** doesn't require this. Dmitri might have strangely high confidences that a non-Heart suit will be drawn supposing that we're working with the standard deck while having strangely low confidences supposing that we're working with the Heartless deck: $Pr_{D,t_i}(D_i|H_1) = Pr_{D,t_i}(S_i|H_1) = Pr_{D,t_i}(C_i|H_1) = .3$ and $Pr_{D,t_i}(D_i|H_2) = Pr_{D,t_i}(S_i|H_2) = Pr_{D,t_i}(C_i|H_2) = .2$. Given the same sequence of evidence as before and referring back to Xenia's original updates, these suppositional confidences would lead to the updates shown in Table 1.

Dmitri and Xenia have diverging confidences here because they don't agree on their suppositional confidences; they interpret the common evidence differently (cf. Glymour 1980, p. 73). One response would then be to require, in order for evidence truly to be "common," that agents also agree on the suppositional confidences associated with their interpretation of the evidence's bearing on an hypothesis – that is, their likelihoods. This added stipulation would certainly take us a step closer to securing convergence.

It's crucial, however, to remember that we're not ultimately aiming at convergence for convergence's sake, but for confidences that are rationally shaped by the world as we experience it. Convergence is viewed as important because it's taken to be indicative of such an *objectification* of otherwise subjectively tainted confidences. But intersubjective convergence of opinion does not itself entail objectification; convergence of opinions need not indicate convergence to confidences rationally shaped by one's evidence or experience of the world. Dmitri and Xenia might agree completely with one another on the likelihoods at issue while agreeing on likelihoods that fail to respond properly to the evidence – for example, if they both accepted Dmitri's likelihoods in the preceding example. In such a case, they would converge to shared confidences that are nonetheless epistemically problematic.[34] It seems that the epistemology of confidence must go beyond **Bayes's Rule** in articulating how our confidences should be constrained by our evidence. These considerations lead us to consider another of Bayes's epistemological postulates.

2.3 Calibration

In developing the solution to his problem, Bayes repeatedly makes use of a second principle, which connects evidence in the form of observed frequencies directly to one's probabilities (Earman, 1992, pp. 27, 52). For example, in Rule 2, Bayes suggests that "if nothing is known concerning an event but that it has happened p times and failed q in $p + q$ or n trials, [...] the probability of its happening in a single trial" lies within some bounds of p/n. The idea is that some evidence, under certain conditions, constrains the confidences that we may rationally have. If all we know about an event type is its observed frequency of occurrence, then our confidence that the event will occur in another trial should equal (or at least lie close to) that frequency. This same idea is echoed by Ramsey (1926, p. 91) in his development of "the logic of truth":

---

[34] Also, recall that agents who agree on assigning prior confidence of 0 to the *true* hypothesis will never converge to the truth by **Bayes's Rule**, even if they converge on some false hypothesis by updating on a common line of evidence, interpreted in the same way using likelihoods that are properly calibrated to evidence per our discussion in §2.3 (Earman 1992, §6.6; Kelly et al. 1997, §2; Howson 2000, p. 210).

Philosophy of Science

> Let us take a habit of forming opinion in a certain way; e.g. the habit of pro-
> ceeding from the opinion that a toadstool is yellow to the opinion that it is
> unwholesome. Then we can accept the fact that the person has a habit of this
> sort, and ask merely what degree of opinion that the toadstool is unwhole-
> some it would be best for him to entertain when he sees it; i.e. granting that
> he is going to think always in the same way about all yellow toadstools, we
> can ask what degree of confidence it would be best for him to have that they
> are unwholesome. And the answer is that it will in general be best for his
> degree of belief that a yellow toadstool is unwholesome to be equal to the
> proportion of yellow toadstools which are in fact unwholesome.

Bayes's presentation of this idea is more careful than Ramsey's in one important respect: he specifies that probabilities are to be identified with known
frequencies "if nothing is known concerning an event but [its frequency]."
Some such clause is needed since it is at best only rational to equate degrees
of confidence with observed frequencies under certain conditions. Indeed, if
in addition to knowing "the proportion of yellow toadstools which are in fact
unwholesome," an agent also knew that a particular yellow toadstool had a
noxious effect on the person who just ingested it, then that agent would surely
be right to have a confidence in its unwholesomeness much greater than the
relevant proportion.

Even so, Bayes's condition of blanket ignorance does not seem to capture the
full range of cases in which such an identification may be safe and reasonable.[35]
Perhaps Bayes is correct that such ignorance is sufficient for using frequencies
to constrain reasonable probabilities, but this condition is not necessary. If an
agent knows the relative frequency in question *and* additionally that a particular
toadstool was plucked from the ground just before twilight, it would still seem
reasonable for the agent to match his or her confidence to that frequency. The
question thus remains: under what exact conditions ought confidences be made
to match relative frequencies?

2.3.1 Calibration and Relative Frequency

Henry Kyburg spent much of his career wrestling with this crucial question, developing Bayes and Ramsey's sentiments into a full-fledged account
of "Direct Inference." He defends the following frequency-based principle of
calibration:[36]

---

[35] Strictly speaking, Bayes's ignorance condition would plausibly rule out application to *all* cases!
There are arguably no actual cases in which an agent knows nothing of an event but that it is
of some salient type that has a relative frequency of occurring (Kyburg, 1977, p. 504). For one
thing, the agent knows that that event is identical to itself, a fact that frustrates simpler accounts
of how to choose an appropriate reference class when seeking an event's relative frequency.

[36] The full account of **Direct Inference** is much more complicated than our brief presentation
suggests. For one thing, there can be instances in which several, equally appropriate reference

Direct Inference. Provided that the relative frequency of events satisfying R within an appropriate reference class of trials satisfying S is known to fall within a specific interval $[r_*, r^*]$, the probability that a *particular* event satisfying S will satisfy R ought to equal that same interval $[r_*, r^*]$.

Importantly, Kyburg allows the "evidential probabilities" arising from **Direct Inference** to be interval-valued. A simple adjustment to the final line of the above principle, requiring the probability to be a member of the salient interval, would allow it to apply to standard real-valued probabilities (Wheeler and Williamson, 2011, §3). However, for Kyburg, since such probabilities are meant to reflect the quality of statistical frequency information available, their level of precision *cannot* exceed the level of precision of the available frequencies.[37] Regardless, in the special cases envisioned by Bayes and Ramsey for which the salient relative frequency is known exactly, this principle agrees with their sentiments in requiring degrees of confidence to be quantitatively precise (equal to $r_* = r^*$).

Developing an example used repeatedly in an important debate between Kyburg and Isaac Levi (Levi, 1977, 1978, 1981; Kyburg, 1977, 1980, 1983), we might wonder whether and how an agent's rational confidence that Petersen was a Protestant ought to be constrained if the agent knows only that *at least* 85% of Swedes living in 1975 were Protestants and that Petersen was such a Swede. So long as the class of Swedes living in 1975 is the appropriate reference class for this problem (more on that in a moment), then **Direct Inference** specifies that the agent's confidence ought to equal the imprecise interval of values [.85, 1].

Of course, there is much that the agent could know about Petersen's specific case that would frustrate this move. The agent might know that Petersen slept in on Sundays, suggesting that this interval is rather too high. Or that Petersen sang tenor in his church choir, suggesting that its lower bound is too low. Such complications are dealt with in **Direct Inference** by the clause specifying that the salient frequency be related to an appropriate reference class. Reichenbach provides a useful starting point: we proceed "by considering the narrowest class for

---

classes give rise to conflicting frequency intervals. Kyburg's full account covers such cases – and many other complications we're glossing over.

[37] **Direct Inference** provides the basis for Kyburg's own unique "evidential" interpretation of probability. For Kyburg, relative frequency is not an appropriate explicandum for probability (Kyburg, 1961, pp. 28–29, 200) – despite the fact that frequencies share in probability's general mathematical structure – but epistemic probabilities are essentially evidentially grounded in frequency information.

which reliable statistics can be compiled" (1949, p. 374). If the agent has "reliable statistics" about Swedes living in 1975 who sing tenor in their church choir,
then that narrower reference class should be used in determining the relative
frequency and consequently the probability that such a Swede is a Protestant.

But how do we determine the appropriate reference class when we fail to
have reliable statistics regarding narrower reference classes? Kyburg responds
that the absence of reliable statistics and objective frequencies amounts to
the absence of constraints on evidential probabilities. Relative to any reference class for which we don't have reliable statistics, an event's evidential
probability will thus be the maximally indeterminate unit interval [0, 1].

This point immediately gives rise to a challenge for Kyburg. Note that there
is something that we know is true of *any* particular event or individual: that
it is identical with that particular event or individual. In addition to knowing
that Petersen is a Swede living in 1975, we also know that Petersen is identical
with Petersen. But then a narrower (indeed, the narrowest) reference class that
Petersen belongs to is the class picked out by the formula "X is a Swede living
in 1975 who is identical with Petersen." If we always plumb for the narrowest
reference class, then it seems that most of Kyburg's evidential probabilities will
be maximally indeterminate.

In response, Kyburg attempts to provide a principled way of choosing
appropriate reference classes such that we balance the amount of information
we know to be true of an individual case ("specificity") with the extent to
which we're allowing the statistical evidence to "constrain our beliefs" ("precision").[38] Instead of always plumbing for narrower reference classes, Kyburg
requires that we refer to *wider* reference classes whenever doing so gives us
strictly more informative, determinate frequency information – that is, whenever the frequency interval relative to any narrower reference class spans the
frequency interval relative to that wider class. We should choose our reference
classes such that they provide statistical evidence that is at once as *specific* to
the individual case in question as possible (i.e., such that they are as narrow
as possible), while still informing our degrees of confidence as *precisely* as
possible.

Returning to the above example, reliable statistics may tell me that at least
85% of Swedes living in 1975 were Protestants. And I may not have the statistics for any narrower reference class of which Petersen is a member (including
the class of which Petersen is the only member). The frequency information
for Protestants among the class of Swedes living in 1975 is [.85, 1]; no frequency information is available relative to any narrower class of which we

---

[38] Kyburg's full account also takes into account a third "richness" desideratum.

know Petersen to be a member, other than the maximally indeterminate fact that the relative frequency is between 0 and 100%: [0, 1]. Since the latter, less precise interval spans the former interval, the appropriate reference class to reference in using **Direct Inference** to determine the evidential probability would be the wider class of Swedes living in 1975. The evidential probability is thus [.85, 1]. Similarly, given reliable statistics demonstrating that Protestants made up between 90% and 95% of the Nordic countries' populations in 1975, **Direct Inference** instructs me to have the more precise evidential probability [.90, .95] based on the wider reference class, effectively neglecting the information that Petersen is a Swede.[39]

Again, the choice to use this class is based on balancing the specifics that we know to be true about Petersen with our desire to let frequencies precisely shape our degrees of confidence. Were we to determine our reference class by bringing to bear all of the information that we know to be true about Petersen, we'd lose any and all precision in our evidential probability (so long as we don't already know whether Petersen is a Protestant); the less precise our knowledge of the frequencies, the less those frequencies are able to inform and constrain our degrees of confidence. As Kyburg and Teng (2001, p. 212) write, "The more complete [specific] body of evidence does not contribute to constraining our beliefs [...] If anything, it merely weakens what we can say."

To the extent that we have reliable statistics about narrower reference classes, however, Kyburg's approach requires us to bring more specific information about Petersen to bear on the problem. In this way, **Direct Inference** guides us in using the available statistics that we do have, potentially relative to wider reference classes, while also implicitly enjoining us to gather more informative statistics related to narrower classes. The idea is to base our degrees of confidence, to the extent possible and available, both in our knowledge of the individual case and in the objective evidence of relative frequencies.

Kyburg's principle of **Direct Inference** – along with its philosophical progeny developed by philosophers such as Pollock (1990), Bacchus (1990), Haenni et al. (2010), and Thorn (2012) – strives to explicate an epistemological constraint on our confidences. The target constraint sharpens the epistemological sentiments of Bayes and Ramsey (among others), who find it compelling that, under certain conditions, one's degrees of belief should be determined by frequency information that he or she has at that time. To achieve a proper

---

[39] Of course, it may be that neither of the frequency intervals (arising from distinct reference classes) spans the other. Kyburg's full account of **Direct Inference** and the determination of evidential probabilities provides guidance in such cases. The reader is directed to Kyburg and Teng (2001, ch. 9) and Wheeler and Williamson (2011) for more extensive treatments.

epistemological fit with the world as we experience it, degrees of confidence plausibly ought to be calibrated, in many cases, to observed relative frequencies.

2.3.2 Calibration and Objective Chance

While this research thus goes some way toward explicating Bayes and Ramsey's thoughts on frequency-probability calibration, there's no reason in principle why we should think that relative frequencies are the only standards worth calibrating to. Indeed, some have argued that there are standards more fundamentally authoritative than frequencies when it comes to calibrating degrees of confidence. David Lewis (1980, pp. 264–265) famously claims that an event type's frequency has great relevance to the probability that one should assign to that event *because* we take it to be evidence of the event's so-called *objective chance* of occurrence. Assume (*per impossibile*?) that I know nothing about an event (e.g., drawing the Ace of Spades from a particular standard deck) except that (1) it has a 1/52 objective chance of occurrence and (2) it has so far occurred in 10% of trials. Lewis maintains that so long as I accept (1), I should have confidence 1/52 that the event will occur in the next trial despite knowing (2).[40] If the observed frequency should have any direct influence at all on my rational confidence in the event, it is only once it (or other evidence) has called into question my beliefs about the event's objective chance.

The upshot, according to Lewis, is that epistemic probabilities should be calibrated first and foremost to objective chances instead of frequencies. Following Levi's (1977) lead, Lewis thus offers the following "Principal Principle" for calibrating degrees of confidence to objective chances:

Principal Principle. If an event is known to have objective chance c of occurring (at time t), the probability (at time t) of that event occurring should equal c, barring any inadmissible evidence.

Following this principle, an agent who knows that a coin is fair should have degree of confidence .5 that the coin will flip heads on any particular toss.

The phrase "barring any inadmissible evidence" in the **Principal Principle** is akin to the requirement in **Direct Inference** that we relativize our frequencies to an "appropriate reference class." There is an acknowledgement that the principle could break down in certain conditions, and thus a clause that bars those

---

[40] Compare to Kyburg (1977, p. 509), who expresses the exact opposite intuition about such cases. Kyburg's response to such cases is influenced by his skepticism about the metaphysical notion of objective chance and the belief that this concept adds little or nothing of practical value beyond what is already provided by relative frequencies (Kyburg, 1976, 2003).

conditions. And just like our discussion of **Direct Inference**, complications arise when we try to clarify the problematic conditions precisely. Of course, inadmissible evidence in the coin case should include the result of the toss in question and similar post-toss giveaways; the probability that the coin tosses heads is clearly not .5 if I observe the coin landing tails, even if I know that the coin is fair! By contrast, admissible evidence arguably includes knowledge of past results and statistical frequencies; given that this coin is fair, the probability that it lands heads on a particular toss is .5, even if I know the results of historic tosses using this coin. Intuitions like these may suffice to give us sufficient conditions under which we may trust the **Principal Principle**; Lewis (1980) himself spends the majority of his defense of the principle spelling out such intuitive, sufficient conditions. However, exactly where to draw the line between admissible and inadmissible evidence – i.e., exactly how to articulate the full conditions under which the **Principal Principle** can be trusted – is a question that remains unanswered.

2.3.3 Discussion

Which calibration principle is better, **Direct Inference** or the **Principal Principle**? If Lewis is right, then the latter principle truly has a *principal* epistemological priority over the other; our probabilities should be calibrated to known objective chances instead of to known frequencies. Still, even if this is the case, **Direct Inference** also has a compelling advantage over the **Principal Principle**. After all, one wants principles that are useful in the actual world. And while we typically have at least some evidence pertaining to an event's chance (often in the form of observed frequencies), we rarely if ever *know* an event's objective chance. To be sure, we may have subjective confidences about various hypotheses pertaining to an event's objective chance, in which case we can calculate the event's *expected* objective chance (Lewis, 1980, p. 267). That's not nothing. However, this move inevitably lands us back into worries about the objective or evidential grounding of our confidences. For our expected objective chances will only be evidentially secured to the extent that the subjective confidences that determine them are. The **Principal Principle** aims to establish an evidential grounding for our confidences, but here we see that it also runs the risk of instead presuming such grounding, at least in the vast majority of cases where evidence leaves us uncertain about objective chances.

In common cases where the best we have (or can acquire) is information regarding an event's relative frequency, the sentiments of Bayes, Ramsey, and Kyburg still seem to motivate rational constraints on degrees of confidence. **Direct Inference** (as explicated by Kyburg and those following in his footsteps) provides our best theory of how to calibrate our confidences to the frequencies we sometimes have and can often discover. Of course, even if we think frequencies can be ignored for objective chances in those cases where the latter are known, we can still affirm a subtle re-articulation of **Direct Inference**. There is no need to choose between the **Principal Principle** and a version of **Direct Inference** that has, as a condition of applicability, that the agent does not have knowledge of the relevant chances. If the agent does have such knowledge, that version of **Direct Inference** no longer applies, and the **Principal Principle** instead kicks in.

Importantly, there is at least one common sort of context in which it's evidently appropriate to apply the **Principal Principle** in spite of uncertainties about chance. When reasoning about the empirical implications of a candidate hypothesis, we're often led to apply a suppositional form of the **Principal Principle**. For example, this form of the principle allows us to say that the probability of a coin landing heads, supposing the coin to be biased .65 towards that outcome, is $Pr(heads|H_{.65}) = .65$. Of course, no knowledge of objective chances is required here, but instead just knowledge of what probabilities would follow on the supposition of salient objective chances. When reasoning under the supposition of such a hypothesis, stipulated chances do indeed seem rightly to constrain our suppositional degrees of confidence.

These observations allow us to disentangle a frequency-based principle (à la Bayes, Ramsey, and Kyburg) from a chance-based principle (à la Levi and Lewis). Both seem to have important roles to play in calibrating epistemically rational degrees of confidence. **Direct Inference** is perhaps the more useful for calibrating non-suppositional confidences, since it only requires statistical information pertaining to actual, relative frequencies. Any non-suppositional application of the **Principal Principle** runs into complications due to the difficulty of "knowing" objective chances; however, the principle applies usefully in theoretical and statistical reasoning when we aim to explore the empirical and experimental implications of hypotheses we are *supposing* to be true (Sprenger and Hartmann, 2019, §12.2).[41]

2.4 Insufficient Reason

Principles of calibration instruct agents on how their confidences should be shaped in the presence of certain types of evidence. By contrast, Bayes's third epistemological postulate instructs agents on what their confidences should be when there is a *lack* of evidence. In the context of his particular problem, Bayes

---

[41] There are still other types of evidence to which it's evidently worth calibrating, including e.g. the confidences of experts or an agent's own future (or past) confidences. For more on such calibration or "deference" principles, see (Pettigrew and Titelbaum, 2014).

(1763, pp. 392–393) discusses "the case of an event concerning the probability of which we absolutely know nothing antecedently to any trials made concerning it." The evidence (or lack thereof) results in a standoff between alternatives, there being "no reason to think that, in a certain number of trials, [the event] should rather happen any one possible number of times than another." Given this lack of reason favoring any one outcome over another, Bayes assigns each possible outcome equal probability.

Pierre-Simon Laplace (1814, pp. 6–7) later famously built this same idea into his very definition of probability. And a century after LaPlace, Keynes (1921, ch. 4) proposes this same idea, in the form of his "Principle of Indifference," as the very cornerstone of his own logical account of probability:

> [I]f there is no *known* reason for predicating of our subject one rather than another of several alternatives, then relatively to such knowledge the assertions of each of these alternatives have an *equal* probability. Thus *equal* probabilities must be assigned to each of several arguments, if there is an absence of positive ground for assigning *unequal* ones.

Adjusting Keynes's articulation of the principle slightly in order to bring out the requisite relation of exclusivity between the propositions in question, let us state this principle of **Insufficient Reason** as follows:

Insufficient Reason. If there is no known reason favoring any one member of a set of mutually exclusive propositions over the others,[42] then we should assign equal probability to each member.

Following this principle, an agent derives substantive confidences from a scant set of evidence. Indeed, an agent can easily be led to quantifiably precise confidences when reason is sufficiently insufficient![43] This happens when the relevant set of alternatives is a partition exhausting the space of possibility. As a simple example, imagine that you encounter a normal-looking six-sided die. If your evidence is sufficiently lacking (so that, e.g., you have no evidence at

---

[42] What exactly is meant by "reason favoring" a proposition? It's tempting to interpret this clause probabilistically – e.g., reason R favors P_1 to P_2 just when $Pr(P_1|R) > Pr(P_2|R)$. But such formally precise characterizations tend to render **Insufficient Reason** empty, amounting to the tautologous norm that we assign equal probabilities to alternatives that have equal probability in light of all we know. Keynes (1921, pp. 53–56) and Carnap (1955, p. 280) both appeal to an intuitive notion of "symmetrical evidence" or "symmetrical knowledge situation" when cashing out this notion.

[43] Granting this point may have bearing on the so-called problem of dilation (§2.2.1). **Insufficient Reason** reveals that amount of evidence can radically come apart from precision of confidences. If, per this principle, lack of evidence can motivate pinpoint precise confidences, is it that strange to think that increases in evidence can sometimes decrease the precision of our confidences? Then again, see the opening discussion of §2.5.

all about historic outcomes of rolling this die), then **Insufficient Reason** may lead you to quantifiably precise, equal confidences that the die will turn up any one of the six possible faces on a single roll: $Pr(S_i) = 1/6$ for $1 \le i \le 6$.

Insufficient Reason seems obviously sound. Indeed, what other confidences could an agent have, in the face of perfectly symmetric or scant evidence, than the equal confidences recommended by this principle? Breaking with **Insufficient Reason** would seem tantamount to favoring alternatives, either purely arbitrarily – "an expression of sheer prejudice" (Jeffreys, 1939, p. 33) – or at best based on considerations that stretch beyond definite reason and the evidence. Either option looks like a paradigmatic case of epistemic irrationality. **Insufficient Reason** thus appears obvious. But such appearances only make this principle's failure all the more spectacular.

2.4.1 Bertrand's Paradox

Insufficient Reason's greatest fault, known now for centuries, is that it ultimately offers inconsistent advice. The probabilities it requires an agent to have in a set of alternatives depend crucially on how the alternatives are described. Equally adequate characterizations of the alternatives inspire equally legitimate, distinct applications of **Insufficient Reason**, across which the same alternative(s) may be assigned different probabilities. This problematic result is often termed "Bertrand's Paradox" after the 19th century French mathematician, Joseph Bertrand, who developed many of its most challenging versions in his *Calcul des Probabilités* (1889).

Consider the following example – of a type suggested not by Bertrand, but by Boole (1854, pp. 369–370), in his discussion of a ball-and-urn case some decades earlier. We're presented with a fair-looking coin. What is the probability of getting nine consecutive heads in our first nine flips of the coin? In the absence of any reason favoring either heads or tails, **Insufficient Reason** requires these mutually exclusive outcomes to have equal probability (of $1/2$) for the first flip of the coin. But how should we apply the principle in order to derive the probability of the first nine flips all coming up heads?

The first step is to describe the possible outcomes with a set of mutually exclusive propositions, but here we have options. First, we could list all possible conjunctions of results from each consecutive flip, providing full detail of the state of affairs we're in with respect to the flips. For example, the state in which only flips three and eight turn up heads would be described with the formal proposition $T_1 \wedge T_2 \wedge H_3 \wedge T_4 \wedge T_5 \wedge T_6 \wedge T_7 \wedge H_8 \wedge T_9$ while the proposition of interest would be formalized as $H_1 \wedge H_2 \wedge \ldots \wedge H_9$. Another option distinguishes alternatives more coarsely by different "structures" of states, listing, for example, the possible numbers of total heads out of the nine flips $\{0H, 1H, \ldots, 9H\}$.

Using the first approach, there are a total of $2^9 = 512$ mutually exclusive outcomes, and reason is equally silent when it comes to favoring any one of these. Applying **Insufficient Reason**, we derive the probability of interest to be $1/512$. By contrast, there are only 10 outcomes described in the second partition. Again, no reason favors any particular number of heads over alternatives, and so **Insufficient Reason** requires $Pr(9H) = 1/10$. The problem, of course, is that $9H$ describes the same outcome as $H_1 \wedge \ldots \wedge H_9$. Thus, **Insufficient Reason** requires us to have radically different degrees of confidence in the same state of affairs. Different descriptions of the same problem reveal that this principle offers inconsistent advice.

A common response is to reject the idea that the two partitionings given above are equally appropriate. One might, for example, point out that there is only one state corresponding to the coin landing heads nine times out of nine, while there are, for example, 9 states in which $8H$ would turn out true and 126 states corresponding to $4H$'s truth. Insofar as one thinks that this comparison provides *reason* to favor $8H$ over $9H$ and even stronger reason favoring $4H$, then it follows that reason does after all favor some members of $\{0H, 1H, \ldots, 9H\}$ over others. Accordingly, **Insufficient Reason** cannot be rightly applied to this partition. Keynes (1921, pp. 59–61) effectively defends the principle in this way by requiring that the alternatives to which it is applied be "ultimate and indivisible": "[**Insufficient Reason**] is not applicable to a pair of alternatives, if we know that either of them is capable of being further split up into a pair of possible but incompatible alternatives of the same form as the original pair." All alternatives, for Keynes, must be equally indivisible; otherwise, the mere fact that some are divisible and others not provides reason to favor the coarser alternatives – in which case, reason is not indifferent: "our knowledge of the *form and meaning* of the alternatives may be a relevant part of the evidence" (p. 61; emphasis in the original).

One obvious problem with this response, however, is that it just seems to assume, not argue for, the appropriateness of the more fine-grained set of alternatives. After all, it only makes sense to think that probability should be assigned to alternatives in proportion to the number of states corresponding to those alternatives, if one already accepts that the states are equally probable. This response thus looks to be begging the question. Other concerns include the fact that **Insufficient Reason** offers inconsistent answers also in cases where alternative partitions are of equal grain to one another (Howson and Urbach, 2006, pp. 271–272). Also, it is easy to refine partitions in such a way that it would seem obviously correct to apply the principle to partitions of coarser, not finer grain (Seidenfeld, 1986, pp. 476–477) – a point that quickly dispels the intuitive case for always applying **Insufficient Reason** to finer partitions.

While intuitions might naively side with applying **Insufficient Reason** to sets of more fine-grained alternatives, there is an intriguing case to be made for siding with coarser alternatives. If one accepts **Bayes's Rule**, then fine-grained state partitions preclude the possibility of learning from experience (Boole 1854, pp. 369–375; Peirce 1883, §10; Keynes 1921, p. 50)! To see this, imagine that the first eight consecutive flips of our coin all land heads. What should our confidence be at this point that the next flip will also land heads? Applying **Bayes's Rule** along with **Conditional Probability**, we have:

$$Pr_{new}(H_9) = Pr_{old}(H_9 | H_1 \wedge \ldots \wedge H_8) = \frac{Pr_{old}(H_1 \wedge \ldots \wedge H_9)}{Pr_{old}(H_1 \wedge \ldots \wedge H_8)}.$$

As we have seen, if we apply **Insufficient Reason** to the partition of states, then $Pr_{old}(H_1 \wedge \ldots \wedge H_9) = (1/2)^9 = 1/512$. Similarly, $Pr_{old}(H_1 \wedge \ldots \wedge H_8) = (1/2)^8 = 1/256$, and thus $Pr_{new}(H_9) = (1/2)^9/(1/2)^8 = 1/2$. Despite the string of common evidence, the agent's confidence remains unshaken that the next flip has a 50/50 chance of coming up heads.

By contrast, if we apply **Insufficient Reason** to our coarser partition (based on possible total numbers of heads), then $Pr_{old}(H_1 \wedge \ldots \wedge H_9) = 1/10$ and $Pr_{old}(H_1 \wedge \ldots \wedge H_8) = 1/9$. Consequently, $Pr_{new}(H_9) = (1/10)/(1/9) = 9/10$. The agent's confidence has been boosted by the string of past, common evidence.

Of course, the same story may be told for any number of flips. After heads turns up in all of the first 999, 999 flips, we might be quite confident indeed that we're not working with a fair coin, but rather that the next toss is almost certain to land heads again. However, if we apply **Insufficient Reason** at the level of states, $Pr_{new}(H_{1,000,000}) = (1/2)^{1,000,000}/(1/2)^{999,999} = 1/2$. Instead applying **Insufficient Reason** to the coarser partition of structures, $Pr_{new}(H_{1,000,000}) = (1/1,000,001)/(1/1,000,000) = .999999$.

Insufficient Reason is supposed to be guiding us in assigning probabilities when we lack evidence. Such probabilities should thus be as uncommitted as possible and open to revision with future evidence. However, by applying **Insufficient Reason** to a partition of state descriptions, we are far from uncommitted and open to revision. On the contrary, such an assignment effectively commits from the start to viewing the coin as fair, and no amount of evidence can shake that initial commitment. Consequently, even after flipping nothing but heads 999, 999 times, agents are instructed to stubbornly maintain their confidences, viewing the coin as unbiased. Alternatively, by using **Insufficient Reason** to spread their probabilities evenly over coarser structure descriptions, agents make no such commitments; their confidences are open to revision, such that even if the coin is indeed fair and flips independent, the relative probabilities of the structures (e.g., $4H$ versus $9H$) eventually reflect that with enough evidence and corresponding revision.

This seems to give us a way forward. If we accept **Bayes's Rule** and presume that we're reasoning in contexts for which we should be able to learn from experience, then we should prefer coarser-grained sets of alternatives to state descriptions when applying **Insufficient Reason**. Carnap (1962) built much of his own logical account of probability and confirmation on the foundations of this observation.

Unfortunately, this response only takes us so far, since it doesn't respond adequately to other versions of Bertrand's Paradox. Here is a well-known example due to van Fraassen (1989, p. 303), but inspired by and very similar to one offered by Bertrand (1889, p. 4) himself: "A precision tool factory produces iron cubes with edge length ≤ 2 cm. What is the probability that a cube has length $l \leq 1$ cm, given that it was produced by that factory?" The obvious answer, derived from **Insufficient Reason**, seems to be $Pr(l \leq 1 \text{ cm}) = 1/2$, since no reason favors the cube's edge length falling into either of the two disjoint and equal intervals $(0, 1]$ and $(1, 2]$. But exactly the same problem can be restated in other ways: A precision tool factory produces iron cubes with side area ≤ 4 cm^2 [or volume ≤ 8 cm^3]. What is the probability that a cube has area $a \leq 1$ cm^2 [or volume $v \leq 1$ cm^3], given that it was produced by that factory? Stating the problem in units of area, **Insufficient Reason** now recommends a probability of $Pr(a \leq 1 \text{ cm}^2) = 1/4$, since no reason favors the cube's side area falling into either of the four disjoint and equal intervals $(0, 1]$, $(1, 2]$, $(2, 3]$, and $(3, 4]$. Applied in the same way to the same problem but stated in terms of volume, **Insufficient Reason** implies $Pr(v \leq 1 \text{ cm}^3) = 1/8$.

Of course, the inequalities $l \leq 1$ cm, $a \leq 1$ cm^2, and $v \leq 1$ cm^3 all convey exactly the same proposition (i.e., the same set of possibilities). Thus, **Insufficient Reason** is once again providing inconsistent results. More importantly, intuitions about learning using **Bayes's Rule** do not in this case favor any of the above partitions over the others. Nor can one defer in this example to any partition of "ultimate and indivisible" alternatives – for this reason, Keynes (1921, pp. 61–62) simply dismissed versions of the paradox such as this from **Insufficient Reason**'s domain of applicability.

2.4.2 Invariance

Another attempt to resolve Bertrand's Paradox attempts to flip the problem on its head. Starting with different articulations of the same identical puzzle, we require from the start that the same puzzle receive identical solutions. Of course, our implicit acceptance of this point is what makes the fact that we *don't* get identical solutions so paradoxical. But we could just as well use this point as a desideratum for assigning probabilities in the first place.

The requirement that different articulations of the very same problem result in the very same probabilities being assigned to alternatives constrains what probability distribution(s) we can assign over the alternatives. The idea is as follows: first, we fish out what we accept to be distinct articulations of the very same problem. By doing so, we reveal differences across which the probabilities in question are invariant. Next, we allow these invariances to constrain, inform, or in some cases even straightforwardly determine how we assign probabilities. The resulting method respects the requisite invariances while arguably also salvaging **Insufficient Reason**. This was the ingenious insight suggested by Harold Jeffreys (1939, §3.10) and developed by E. T. Jaynes (2003).

Let's return to van Fraassen's perfect cube factory in considering this attempted resolution. We start with our acceptance of the claim that the three distinct articulations in fact describe the same identical problem. They differ with respect to what scale they use to articulate the problem (cm, cm^2, or cm^3), but this choice of scale should not make a difference to the probabilities of alternatives. Accordingly, we require that the solution to this problem be *scale-invariant*; probabilities of alternatives should not depend on the scale we use to describe those alternatives.

It turns out that this requirement by itself already determines for us a unique probability distribution! In place of the problematic uniform probability distribution used in the paradoxical version of the problem, the desideratum of scale-invariance implies that we should use a *log-uniform* distribution. Such a distribution assigns equal probabilities to intervals $[x, y]$ with equal measure $\log y - \log x$.

To apply this distribution in our example, we must slightly alter the details and explicitly require all possible cubes to have non-zero (positive) edge lengths – this is necessary to sidestep the fatal logarithm of zero (van Fraassen, 1989, pp. 309–310). Let the cubes produced by the precision tool factory have edge length between 1 and 3 cm inclusive. Then the three questions that ought to receive the same answer are:

- What is the probability that a cube has side length $l \leq 2$ cm, given that it's between 1 and 3 inclusive?
- What is the probability that a cube has side area $a \leq 4$ cm^2, given that it's between 1 and 9 inclusive?
- What is the probability that a cube has volume $v \leq 8$ cm^3, given that it's between 1 and 27 inclusive?

For the first question, out of unit measure $\log 3 - \log 1$, the cases in which $l \leq 2$ have measure $\log 2 - \log 1$; thus:

$$Pr(l \leq 2) = \frac{\log 2 - \log 1}{\log 3 - \log 1} = \frac{\log 2}{\log 3} = .631$$

The second and third questions are similarly solved and lead to the desired, identical answer:

$$Pr(a \leq 4) = \frac{\log 4 - \log 1}{\log 9 - \log 1} = \frac{\log 2^2}{\log 3^2} = \frac{2 \log 2}{2 \log 3} = \frac{\log 2}{\log 3} = .631,$$

$$Pr(v \leq 8) = \frac{\log 8 - \log 1}{\log 27 - \log 1} = \frac{\log 2^3}{\log 3^3} = \frac{3 \log 2}{3 \log 3} = \frac{\log 2}{\log 3} = .631.$$

This solution escapes paradox, assigning the same probabilities to the same alternatives across different articulations of the same problem. Note that **Insufficient Reason** is still being applied, since we continue to assign equal probabilities to intervals that reason does not distinguish. But the invariance relation provides us essential, additional guidance in determining the precise such intervals that must be used in any sensible solution to this problem – in this case, intervals $[x, y]$ with equal measure $\log y - \log x$.

Just how successful a strategy this is for avoiding Bertrand's Paradox remains a matter of continuing debate (Seidenfeld 1979; Jaynes 2003, ch. 12; Howson and Urbach 2006, §9.a.3; Williamson 2010, ch. 9, Williamson 2017, §8.1). One considerable problem is that there can be more, equally compelling invariances at play, which come into conflict with each other. To see this, note that we may plausibly require that the probabilities in our previous example are not only invariant over different scales but also translation invariant – that is, invariant over a constant shift in the range of possible side lengths for cubes coming out of our factory. That is, we might plausibly require that all three of the following questions have the same answer:

1. What is the probability that a cube has side length $l \leq 2$ cm, given that it's between 1 and 3 inclusive?
2. What is the probability that a cube has side area $a \leq 4$ cm^2, given that it's between 1 and 9 inclusive?
3. What is the probability that a cube has side length $l \leq 3$ cm, given that it's between 2 and 4 inclusive?

1 and 2 are the same question on different scales, and 3 involves a translation of scale from 1. However, the only distribution that provides the same answer to questions 1 and 2 (the log-uniform distribution) renders a different answer to question 3: $Pr(l \leq 3) = .585$ when l's range is $[2, 4]$.

The upshot is that there is no probability distribution that allows us simultaneously to respect the desired scale- *and* translation-invariance of this problem. In cases like this, then, one must prioritize certain desirable invariances over others. Thus, the fundamental concern highlighted by Bertrand's Paradox is still pressing. In some cases, there are different ways to describe and partition the alternatives, each partitioning possibly determined by a desirable set of invariances. But the probability of some one alternative may be sensitive to the partitioning that one chooses.

2.5 MAXENT

Let's consider one more epistemological postulate, which relates the ideas we've discussed in interesting ways. This principle can be motivated by considering the following. **Insufficient Reason** aims to direct our confidences in the absence of telling evidence or reason. Principles of calibration such as **Direct Inference** and the **Principal Principle**, by contrast, legislate what confidences we ought to have in the presence of particular sorts of evidence. But what should our confidences look like in situations that are somewhere in between, that is, in situations to which calibration principles apply, but where the (frequency- or chance-based) evidence does not suffice to calibrate our confidences with quantifiable precision?

In such cases, we've seen that Kyburg's principle of **Direct Inference** requires that calibrated confidences be no more precise than the learned frequency information.[44] That is, Kyburg's account responds to these in-between cases by requiring that our confidences be no more precise than the evidence. A great virtue of this response is its seemingly straightforward motivation on evidentialist grounds. Epistemically rational confidences should only be constrained by evidence. Once the evidence has calibrated our confidences to its full extent then, any further precision in our confidences cannot be inspired by the evidence and thus cannot be epistemically warranted.

But another possible response avoids the complication of introducing "imprecise probabilities" while arguably still respecting this evidentialist motivation. This response makes use of an appealing extension to **Insufficient Reason**. If **Insufficient Reason** leads one to quantifiably precise confidences when there is no evidence that will do the job, a plausible extension of this principle might do the same for evidence that only partially does the job. This

---

[44] And it's easy to imagine a similar version of the **Principle Principal** requiring, for example, that if an event is known to have objective chance $c \geq .95$, then barring any inadmissible evidence (including information giving us more precise knowledge of the objective chance), the agent's confidence in that event occurring should be no more precise than $Pr \geq .95$ (or the set $\mathcal{P} = [.95, 1]$).

is the motivation behind Jaynes' development of what he calls "the entropy principle" (1957; 2003, ch. 11). Once we have let the evidence constrain our confidences in accordance with calibration principles, we want confidences that commit no further. That is, these confidences should be as uncertain, equivocal, and unbiased as possible, within the constraints set by the evidence. Shannon (1948, §6) famously proves that the following measure of the "amount of uncertainty" represented in a discrete probability distribution uniquely (up to choice of logarithmic base) satisfies a set of plausible conditions of adequacy:

$$H(Pr) = - \sum_{i=1}^{n} Pr_i \log(Pr_i).$$

$H(Pr)$ is often referred to as the relevant distribution's measure of *entropy*.[45] Jaynes (2003, p. 351) refers to H as "a fundamental measure of how uniform a probability distribution is," since this measure increases as Pr approaches a uniform distribution and takes a maximal value, relative to any discrete partition, when all members of the partition are assigned equal probability. Note that **Insufficient Reason**'s prescription to assign equal probabilities thus corresponds to the advice to *maximize entropy* (alternatively, *minimize bias*, *minimally commit*, or *maximally equivocate*). Jaynes's generalization of this advice takes the form of the following principle:

MAXENT. Given evidence E and partition Ω, accept the probability distribution over Ω that maximizes $H(\cdot)$ while satisfying the constraints imposed by calibrating on E.

Having no knowledge of the objective chance that Peterson, a particular Swede living in 1975, was Protestant but equipped with the statistical knowledge that at least 85% of Swedes in 1975 were Protestant, **Direct Inference** directs us to the imprecise interval $[.85, 1]$ that Peterson was Protestant. As we've seen, this principle of calibration tells us to stop there, having confidences that are no more precise than our positively constraining, statistical evidence. By contrast, **MAXENT** has us *start* there, requiring that we adopt a quantifiably precise confidence out of the calibrated interval. Given the partition of religious affiliations in Sweden at the time {*Protestant*, Catholic,

---

45 $H(Pr)$ of course only applies in the discrete case. Shannon's (1948, §20) parallel function for the continuous case is (where Ω is x's continuous range of values, and f is Pr's probability density function):

$$H^*(Pr) = - \int_{x \in \Omega} f(x) \log f(x) dx.$$

Orthodox, Other Religions, Unaffiliated}, **MAXENT** instructs us to come as close as possible to the associated uniform distribution {.2, .2, .2, .2, .2} while staying within the bounds of the relevant calibrated interval. Having no other relevant statistical information in this case, the result is the distribution that assigns precise confidence .85 to Peterson being Protestant and precise confidence .0375 to Peterson being any one of the remaining four options.

When E provides no reason in favor of any member of Ω over any other, calibration on E does not rule out a uniform distribution. In such a case, **MAXENT** recommends the uniform distribution and so amounts to **Insufficient Reason**. However, **MAXENT** additionally speaks to our target cases in which E provides reason in favor of certain alternatives, ruling out the uniform distribution as it were. In this case, **MAXENT** still applies and instructs us to choose the distribution that maximizes entropy out of those (nonuniform) distributions still standing after E has had its say. Our original question was what we ought to do in situations where calibration principles apply, but where the (frequency- or chance-based) evidence does not suffice to calibrate our confidences with quantifiable precision. **MAXENT**'s answer is that we should calibrate with as much precision as the evidence warrants and then adopt the precise probability distribution out of the remaining alternatives that maximizes entropy. The confidences this method guides us to are quantifiably precise. However, the fact that they are singled out ultimately by **MAXENT** implies that they are, in a very exact sense, minimally committed beyond the constraints of the evidence.[46]

MAXENT provides a bridge between calibrated confidences and intuitions about what to do when the evidence is not fully sufficient for pinning our confidences down – intuitions that motivate **Insufficient Reason**. But **MAXENT** also relates in fascinating ways to **Bayes's Rule** and principles for updating confidences. Specifically, different applications of **MAXENT** relative to changing evidence sets across time constitute a straightforward means of updating:

MAXENT Updating. Given a change in total evidence from E_{old} to E_{new} and partition Ω, an agent should (a) initially accept that probability distribution over Ω which maximizes $H(\cdot)$ while being calibrated to E_{old}, and (b) then update to that distribution over Ω which maximizes $H(\cdot)$ while being calibrated to E_{new}.

A key result proved by Seidenfeld (1986, Result₁ and Corollary, pp. 471–472) demonstrates that **MAXENT Updating** is actually equivalent to

---

[46] Williamson (2007) provides a thorough defense of **MAXENT** as a preferable approach to Kyburg's imprecise evidential probabilities.

Bayes's Rule, but only under a particular set of substantive conditions. While this point is often assumed to reflect poorly on **MAXENT Updating** (Shimony, 1985; Skyrms, 1985; Seidenfeld, 1986), there are certainly some things to be said in favor of this rule.

For one thing, **MAXENT Updating** easily avoids the two complications facing **Bayes's Rule** highlighted in §2.2.2 (Williamson, 2010, p. 84). Unlike **Bayes's Rule**, **MAXENT Updating** doesn't entail that newly learned evidence is learned with absolute certainty (i.e., takes unit probability); indeed, one central reason why principles of calibration may often lead to imprecise confidences is because the evidence is uncertain. Also, **MAXENT Updating** allows agents rationally to become increasingly uncertain of previously accepted evidence or forget some evidence altogether. In such a case, we simply proceed as usual: we take the new evidence (in whatever ways it has evolved), recalibrate, and then maximize entropy. Finally, it should be noted that Williamson argues against the idea that **MAXENT Updating**'s departure from **Bayes's Rule** is a mark against the former principle. Williamson (2010, §4.2; 2011, §4) argues systematically, to the contrary, that in every case in which the two principles may diverge, **MAXENT Updating** "is a more appropriate form of updating than [Bayes's Rule]."

Our discussion throughout this section has raised more questions than it has answered. But alas, epistemology is a branch of philosophy, and such is philosophy! If there can be such a thing as a complete, sensible Bayesian epistemology of confidences, we are unsurprisingly still very much in the process of developing it. Nonetheless, in the next section, we will put this unfinished product to use by exploring a few questions in the epistemology of scientific reasoning from the Bayesian perspective.

3 Scientific Reasoning

§1.1 and §1.2 contrasted the truth-functional approach taken in classical logic to the logic of uncertainty provided by probability theory. Additionally, we suggested reasons to think that a probabilistic approach to logic (and epistemology) might be particularly well-suited for studying scientific reasoning. Scientific research is permeated with uncertainty. This shouldn't come as a surprise; science is about advancing the frontiers of empirical knowledge, not about stating and restating tired certainties. But if scientists are often reasoning about the world in the face of deep uncertainties – testing bold conjectures, entertaining theoretical speculations, considering competing explanations, and so on – then it would seem obvious that we should turn to a logic and epistemology of uncertainties in order to shed light on what they're doing.

The truth-functional semantics of deductive logic has, of course, been applied to the study of scientific reasoning. But in light of the pervasiveness of uncertainty in science and the availability of a probabilistic approach that deals directly in uncertainties, the deductive approach seems crude and unrefined. At least prima facie, a logic and epistemology of uncertainties provides a more appropriate approach to studying the uncertainty-laden contexts of scientific reasoning. At any rate, this final section will draw upon some of the resources developed in Sections 1 and 2 in order to see whether these tools help shed light on certain topics and questions of scientific reasoning.

Center stage throughout much of our discussion will be **Bayes's Theorem** and the logical relations that it may be used to highlight between hypotheses and evidence. Previously in this Element, we have often used Greek letters such as ϕ and ψ as metavariables standing in for any proposition in some Pr's domain. Turning our attention now specifically to the study of scientific reasoning, we will often want to denote hypotheses, statements of evidence, and an agent's background beliefs generically. Let's use lower-case h, e, and k for these purposes (sometimes with numerical subscripts) – since we use capital letters to stand for particular propositions. The common statement of **Bayes's Theorem** in philosophy of science then is as follows:

$$Pr(h|e \wedge k) = \frac{Pr(h|k)Pr(e|h \wedge k)}{Pr(e|k)}.$$

The following terminology – inspired in part by the typical Bayesian endorsement of **Bayes's Rule** (§2.2) – has by now become standard:

- $Pr(h|e \wedge k)$: The *posterior probability* of h.
- $Pr(h|k)$: The *prior probability* of h.
- $Pr(e|h \wedge k)$: The *likelihood* of h, or the *likelihood* of e given h.
- $Pr(e|k)$: The *expectedness* of e.

Bayes's Theorem clarifies that an h's posterior probability $Pr(h|e \wedge k)$ is an increasing function of its prior probability and likelihood and a decreasing function of e's expectedness. Each of these observations has an intuitive reading. First, one ought to be confident about h, in the light of e, to the extent that one was already confident in h prior to considering e, and to the extent that e is likely given h. Second, insofar as evidence e was already expected to be true, it has a diminishing effect on the posterior probability of h. Along these same lines, note that if e is not made any more or less probable by supposing h, $Pr(e|h \wedge k) = Pr(e|k)$, then h's probability will not be affected by e, $Pr(h|e \wedge k) = Pr(h|k)$.

3.1 Confirmation

In this section's investigations, we'll evaluate some modes of scientific reasoning by asking specifically whether they do anything to warrant an increased confidence (alleviation of uncertainty) in a target scientific conclusion or hypothesis – and if so, under what conditions. Where probabilities are interpreted as rational degrees of confidence, this amounts to investigating the conditions under which particular scientific methods work to increase the probability of some target conclusion or hypothesis.

We'll thus follow a general trend in Bayesian epistemology of science and center our evaluations and investigations on the notion of "incremental confirmation." An hypothesis h is incrementally confirmed by some evidence e, in the light of an agent's background k, exactly when e increases h's probability, conditional on k: $Pr(h|e \wedge k) > Pr(h|k)$. It's easy to misconstrue this central notion of confirmation in several different ways; thus, let's set up some signposts to help us not lose our way.

First, it's easy to read the statement "*e* confirms *h* relative to *k*" as meaning that, in the light of k, evidence e *proves* that h must be true, or at least renders h probably true beyond reasonable dispute. In contemporary terms, this would be to mistake our notion of incremental confirmation for an *absolute* sense of confirmation (Fitelson, 2001b). Following Carnap's (1962, p. xvi) lead, let's distinguish the concepts probabilistically:[47]

Absolute Confirmation. h is confirmed (on the basis of e and k) if and only if $Pr(h|e \wedge k) > t$, where t is a suitable threshold signifying a value sufficiently high to put h beyond reasonable dispute.

Incremental Confirmation. e confirms h (to some degree), relative to k, if and only if $Pr(h|e \wedge k) > Pr(h|k)$.

Absolute Confirmation is meant to be taken holistically, encapsulating the bearing of *all* of an agent's evidence (including all relevant evidence found in the background term k) to h – Carnap (1947, 1962) explicitly requires this in his principle of "total evidence." **Incremental Confirmation**, by contrast, spells out a more targeted notion of confirmation pertaining to whether a distinct bit of evidence e confirms h in the light of an agent's background k. That these two notions of confirmation can easily come apart is made obvious by the following two possibilities:

Absolute but not Incremental. $Pr(h|k) > Pr(h|e \wedge k) > t$.
Incremental but not Absolute. $t > Pr(h|e \wedge k) > Pr(h|k)$.

---

[47] Carnap refers to **Absolute Confirmation** and **Incremental Confirmation** as "firmness" and "increase in firmness" respectively.

In the first case, h is (absolutely) confirmed in light of e and k while simultaneously being (incrementally) disconfirmed by that same e conditional on the same k; that is, h remains sufficiently probable ($> t$) given e and k despite the fact that e has decreased its probability. In the second case, while e (incrementally) confirms h, it still leaves h overall (absolutely) unconfirmed since it does not raise h's probability over threshold t.

In everyday life, we often use the language of confirmation in the absolute sense.[48] Nonetheless, it's purely the incremental notion that we'll make use of in this section. Additionally, we'll employ the following corresponding notions of incremental disconfirmation and confirmational irrelevance:

Incremental Disconfirmation. e *disconfirms* h (to some degree), relative to
k, if and only if $Pr(h|e \wedge k) < Pr(h|k)$.
Confirmational Irrelevance. e is confirmationally irrelevant to h, relative to
k, if and only if $Pr(h|e \wedge k) = Pr(h|k)$.

Intuitively, evidence confirms an hypothesis exactly when (and to the extent that) it increases the probability of that hypothesis. Evidence disconfirms an hypothesis to the extent that it makes that hypothesis less likely. And evidence is irrelevant to an hypothesis when it neither confirms nor disconfirms it.

Second, it's easy to read the statement "e confirms h (relative to k)" as a purely logical and objective statement about the evidence-hypothesis(-background) relation. But this can be a mistake for various reasons. For one thing, we must keep in mind the agent-relative nature of our interpretation of probability. We interpret probabilities as the confidences of an idealized agent (§1.4.1). Different such agents correspond to different probability distributions over a salient domain of propositions, and in some cases, it can turn out that $Pr(h|e \wedge k) > Pr(h|k)$ is satisfied by some but not all of these distributions. That is, even holding k fixed, whether e confirms h may differ between idealized agents. Postulates we've explored from Bayesian epistemology (Section 2) may go further and preclude all idealized confidences for which the relevant confirmation relation doesn't hold; but even in such cases, h's confirmation of e would at best hold as a matter of *epistemological* – and not purely *logical* – course.

Our bridge principle of **Consistency** (§1.4.2) requires, in order for an actual agent's confidences to cohere logically, only that their formalizations be jointly entailed by at least one single probability distribution. Confidences (be they

---

Although not too much should be made of this. Empirical research by Tentori et al. (2007)
supports the idea that we also often reason in terms of confirmation in the incremental sense.

quantifiable, qualitative, relative, etc.) prescribed purely via the logic of **Consistency** then are those that all probability distributions agree upon, being implied by the probability axioms – else one could hold opposing confidences and still satisfy **Consistency** by reference to one of the dissenting distributions.

Whether a claim about confirmation expresses a purely logical relation between h, e, and k will thus depend on the details of a case. An example in which this plausibly is the case is provided by Bayesian confirmation theory's ability to accommodate Popper's (1959, pp. 9–10) deductive notion of falsification. For Popper, $\neg e$ decisively falsifies h ("with the help of" some k) if $h \wedge k$ entails e and then we come to discover $\neg e$ (where k alone has not already falsified h). What does the Bayesian approach say about $\neg e$'s confirmatory impact on h here? Since the deductive entailment relation requires that $Pr(e|h \wedge k) = 1$, $Pr(\neg e|h \wedge k) = 0$ and the answer is easily derived:

$$Pr(h|\neg e \wedge k) = \frac{Pr(h|k)Pr(\neg e|h \wedge k)}{Pr(\neg e|k)} = 0.$$

That is, the Bayesian concludes here that, as a matter of logical fact, h is indeed decisively disconfirmed by $\neg e$ (in the light of k).

A slightly more complicated example is provided by Bayesianism's accommodation of the Hypothetico-Deductive (Hempel, 1945, p. 98) requirement that an hypothesis h be confirmed by evidence e it deductively entails – possibly with the help of background ("auxiliary") information k, where k doesn't already entail e on its own.[49] As a purely logical matter of fact, the Bayesian can demonstrate the weaker conclusion that there is no disconfirmation in such a case:

$$Pr(h|e \wedge k) = \frac{Pr(h|k)Pr(e|h \wedge k)}{Pr(e|k)},$$

$$\therefore \frac{Pr(h|e \wedge k)}{Pr(h|k)} = \frac{Pr(e|h \wedge k)}{Pr(e|k)} = \frac{1}{Pr(e|k)} \geq 1,$$

$$\therefore Pr(h|e \wedge k) \geq Pr(h|k).$$

But in order to make the stronger claim that e confirms h, we have to go beyond logic and draw upon epistemological resources. $Pr(h|e \wedge k) > Pr(h|k)$ follows specifically from the additional premise that e is uncertain given k, $Pr(e|k) < 1$. But this premise doesn't follow from the probability calculus itself. Accordingly, to argue that e confirms h, we must supply an epistemological case for thinking that an agent ought not have certainty in e, suppositional on k alone.

---

[49] To take a common (if simplistic) example, Boyle's Law may be confirmed by observing a gas in a closed system undergo a particular change in volume, since it entails such an occurrence when combined with auxiliary information about that gas's change in pressure and temperature.

Such a case might, for example, plausibly be built upon the epistemological principle of **Regularity** (§2.2.2).

The upshot is that, in many cases, we will need to go beyond the pure logic of consistency in order to argue that a target confirmation relation holds. But exploring whether this is true in different cases will prove to be an important part of our task. In general, in cases for which $Pr(h|e \wedge k) > Pr(h|k)$ isn't implied by the probability calculus itself, we may uncover additional conditions under which the relevant mode of scientific reasoning *is* demonstrably confirmatory. Such an investigation sheds important light on the usefulness and limitations of the scientific methods in question.

The remainder of this section exemplifies the Bayesian approach to the logic and epistemology of scientific reasoning, focusing on three main topics: (1) explanatory reasoning, (2) robustness analysis (including brief discussions of evidential diversity and hypothesis competition), and (3) simplicity and evidential fit. Within any one of these topics, we'll focus on a particular Bayesian approach one might take out of potentially many. Each topic merits a book-length treatment unto itself; thus, our exploration in each case will be cut short, leaving many more questions and potential criticisms standing than answered. However, the hope is that our admittedly idiosyncratic choice of topics results in a collection of discussions that inspire the reader to look more deeply into each issue and to explore more widely the world of Bayesian investigations into scientific reasoning.

3.2 Explanatory Reasoning

In advancing the frontiers of our understanding of the empirical world, scientists obviously do not rest content merely cataloguing their observations of what is the case. If anything, such observations rather provide an important starting place for scientific inquiry, which presses further for explanations of *why* things are the way we observe them to be. Philosophers of science throughout the ages have remarked on the centrality of explanation to scientific inquiry. Hence, Aristotle begins his *Physics* with the declaration, "Knowledge is the object of our inquiry, and men do not think they know a thing till they have grasped the 'why' of it" (194^b 17–20, as translated in Barnes 1984). Immediately after making this statement, Aristotle presents his doctrine of the four causes, an early theory of scientific explanation at the core of his account of how we are to investigate the natural world. In modern times, Einstein and Infeld (1938) present the history of physics as a "great mystery story" in which "at every stage we try to find an explanation consistent with the clews already discovered" (p. 4). At the start of their classic investigation into "the logic of explanation," Hempel and

Oppenheim (1948, p. 135) express the same sentiment: "To explain the phenomena in the world of our experience, to answer the question 'why?' rather than only the question 'what?', is one of the foremost objectives of all rational inquiry."

A philosophy of scientific reasoning must accordingly seek an understanding of the nature and role of explanation. Philosophers of science have traditionally placed most of their collective focus inquiring about the very nature of scientific explanations. However, we'll take a different tack here, focusing instead on the common epistemic impact that many explanations (be they causal, unificatory, nomic, mechanistic, etc.) have on reasoners considering and evaluating them.[50] That is, we'll be sidestepping investigations into the properties that qualify a hypothesis as a potential explanation of some explanandum, or the exact conditions under which an explanation qualifies as a scientific explanation. We'll stipulate that the hypotheses we have in mind do indeed potentially explain their explananda, according to a satisfactory account of the nature of explanation. And we'll then more narrowly target the epistemology of explanation by, for example, exploring epistemic conditions under which reasoners consider such potential explanations to be good or powerful.

3.2.1 Peirce and Power

This was the approach to the study of explanation's role in scientific reasoning taken by C. S. Peirce. Peirce was not silent on the nature of explanation; indeed, in various places (e.g., 1935, §1.89; 1893; 1958, §7.192), he offers an account of explanation that is a clear precursor to Hempel's (1965) "Deductive-Nomological" and "Inductive-Statistical" models.[51] However, Peirce recognizes a legitimate question of explanation's epistemological function distinct from the question of explanation's definition.[52] In the sixth and final

---

[50] A more thorough defense of this move, focusing on the epistemology of explanation without already having in hand a satisfactory account of the nature of explanation, may be found in (Schupbach, 2011b, §1.2).

[51] Thanks to Jonathan Livengood for a clarifying discussion on this point.

[52] This is true even recognizing that Peirce's theory of explanation and epistemology of explanation are in fact intimately tied together. E.g., Peirce (1958, 7.199) writes:

"We notice a plant that is flagging on a hot summer day: next morning it stands up again fresh and green. 'Why has it revived in the morning?' – 'Oh they always do'." One may smile at the naïveté of this; and certainly, it is not an explanation in the proper sense of the word. Still, its general function is the same as that of explanation; namely, it renders the fact a conclusion, necessary or probable, from what is already well known.

installment of his "Illustrations of the Logic of Science," Peirce (1878, p. 472) suggests a style of inference via which reasoners adopt explanatory hypotheses: "[W]e find some very curious circumstance, which would be explained by the supposition that it was a case of a certain general rule, and thereupon adopt that supposition." A quarter of a century later, in the last of his seven Harvard lectures (delivered in 1903), Peirce (1935, 5.189) offers a fuller account of this explanatory inference:[53]

> Long before I first classed abduction as an inference it was recognized by logicians that the operation of adopting an explanatory hypothesis – which is just what abduction is – was subject to certain conditions. Namely, the hypothesis cannot be admitted, even as an hypothesis, unless it be supposed that it would account for the facts or some of them. The form of inference, therefore, is this:
>
> > The surprising fact, e, is observed;
> > But if h were true, e would be a matter of course;
> > Hence, there is reason to suspect that h is true.

Peirce suggests that explanatory hypotheses (i.e., those that do offer potential explanations of their corresponding explananda) have the effect of making their explananda more expected (or less surprising). Let's call this alleged explanatory virtue "power" (McGrew, 2003). While we will eventually call into question the generality of Peirce's schema for explanatory inference, the notion of power does seem to capture one very common epistemic mark of explanations. Indeed, when scientists comment on the traits of a good scientific explanation, they often describe the explanation's ability to render expected the otherwise unexpected. To cite just a couple examples, throughout the *Origin*, Darwin (1872, e.g., on pp. 626–627) repeatedly identifies the explicability of some fact with it ceasing to be strange and being expected or anticipated in the light of his theory of Natural Selection. Einstein and Infeld (1938, p. 4) compare the process of discovery in the history of physics to the situation in a mystery novel when facts that were previously "strange, incoherent, and wholly unrelated" suddenly become "certain" in the light of the powerful explanation.

In working up to a Bayesian explication of this notion, let's first generalize Peirce's informal idea, stating the salient notion of power *as a matter of degree*: an explanatory hypothesis has power over its explanandum *to the extent that* it increases the expectedness of that proposition. So, for example, a geologist will favor a prehistoric earthquake as a powerful explanation of observed

---

[53] For the sake of notational consistency, I've replaced Peirce's "A" and "C" with "h" and "e" respectively.

deformations in layers of bedrock to the extent that deformations of that particular character, in that particular layer of bedrock, and so on are more expected given the occurrence of such an earthquake. Unlike Peirce's articulation of abduction, this condition allows that an hypothesis may provide a powerful potential explanation of a surprising proposition and still not render it a matter of course in any sense. Additionally, our condition does not suggest that a proposition must be surprising in order to be explained; a hypothesis may make a proposition much less surprising (or more expected) even if the latter is not so surprising to begin with.

The idea that some h has power over e to the extent that it increases e's expectedness leads to further implications pertaining to power.[54] Among other things, we may point out the following:

1. h has its maximal degree of power over e exactly if it maximally increases e's expectedness, rendering e completely certain; by contrast, h has its *minimal* degree of power over e exactly if it maximally decreases e's expectedness, rendering e certainly false.
2. We might also straightforwardly define a notion of *negative* power as corresponding to the extent to which an h renders e *less* expected (or more surprising). Of course, complete lack of power (positive or negative) then corresponds to the case in which h does nothing to increase or decrease the degree to which we expect e.
3. The more expected h makes e, the less expected it makes e's falsity; or, more directly stated in terms of power, the more power h has over e, the less it has over $\neg e$.

3.2.2 Measuring Explanatory Power

We are now ready to apply the Bayesian machinery in order to explore the logic and epistemology of explanatory power. The goal is for this normative account to shed light on the merits and limitations of explanatory reasoning in terms of power. Here, we'll assume that neither the explanandum e nor the explanatory hypothesis h are assigned extreme probabilities. This reflects the uncertainty of h inherent in contexts of explanatory reasoning about h as well as the Peircean idea that e is, to some degree, unexpected (at least apart from h) in such contexts.

---

54 When we talk about h "increasing e's expectedness" here, we're using this phrase loosely. Strictly speaking, as noted in the first pages of this section, e's expectedness refers to the term $Pr(e|k)$. By h increasing e's expectedness, we mean that e's expectedness conditional on h (i.e., the likelihood of e given h) is higher than e's expectedness: $Pr(e|h \wedge k) > Pr(e|k)$.

The key move in order to cast explanatory power probabilistically is to iden-
tify degrees of expectedness with the degrees of confidence that Bayesians
associate with probabilities. Doing so allows us to formalize the above observations as conditions of adequacy for any Bayesian measure of explanatory
power $\mathcal{E}$ (Schupbach and Sprenger 2011, pp. 109–113; Cohen 2015, pp.
98–101; Schupbach 2017, pp. 44–45; Sprenger and Hartmann 2019, §7.3).
Such conditions include the following:[55]

Positive Relevance. For any fixed value of $Pr(e)$, the greater the degree of
    statistical relevance between e and h, $Pr(e|h) - Pr(e)$, the greater the
    value of $\mathcal{E}(e, h)$.
Maximality. $\mathcal{E}(e, h)$ is maximal if and only if $Pr(e|h) = 1$.
Neutral Point. $\mathcal{E}(e, h) = c$ if and only if $Pr(e|h) = Pr(e)$, where c is a
    constant.
Symmetry. $\mathcal{E}(e, h) = -\mathcal{E}(\neg e, h)$.
Irrelevant Conjunction. If h_2 is independent of e, h_1, and $e \wedge h_1$, then
    $\mathcal{E}(e, h_1 \wedge h_2) = \mathcal{E}(e, h_1)$.

Positive Relevance straightforwardly explicates the basic sentiment that h
has power over e to the extent that e becomes more expected (probable) in the
light of h. Similarly, **Maximality** and **Neutral Point** associate $\mathcal{E}$'s maximal and
"neutral" values with the case of h rendering e maximally expected (probable)
and leaving e's expectedness unaffected respectively. **Symmetry** is motivated
by idea 3 above; in short, e is more expected precisely to the extent that $\neg e$ is
less expected.[56] As Cohen (2015, p. 101) points out, if one accepts **Symmetry**,
then $c = 0$ in the statement of **Neutral Point**.[57] Finally, interpreted in the
light of **Neutral Point**, **Irrelevant Conjunction** amounts to the assertion that
$h_1 \wedge h_2$ will have precisely the same amount of power over e as does h_1 alone
(i.e., they increase the expectedness of e to exactly the same extent) in the case
that h_2 has no power over e, h_1, or any logical combination of e and h_1.

Such formal conditions are useful in determining a uniquely satisfactory
Bayesian measure of power. Specifically, **Maximality, Neutral Point, Symmetry**, and **Irrelevant Conjunction** jointly determine the following measure

---

[55] As we've seen throughout this Element, the background term k always belongs to the right
of the solidus "|" in Bayesian formalizations. Nonetheless, here and in the remainder of this
section, I choose for the sake of readability to leave k implicit in all formalizations.

[56] As Robert Hartzell helpfully points out (personal correspondence), **Symmetry** can also be motivated as a means to rule out unfalsifiable explanations of everything, since it precludes h from
having positive explanatory power both over any e and its negation $\neg e$.

[57] If $Pr(e|h) = Pr(e)$, then $Pr(\neg e|h) = 1 - Pr(e|h) = 1 - Pr(e) = Pr(\neg e)$. Thus, by
Neutral Point, if $Pr(e|h) = Pr(e)$, then $\mathcal{E}(e, h) = \mathcal{E}(\neg e, h) = c$. But by **Symmetry**,
$\mathcal{E}(e, h) = -\mathcal{E}(\neg e, h)$. Thus, $\mathcal{E}(\neg e, h) = -\mathcal{E}(\neg e, h) = c = 0$.

as uniquely best, assuming that our measure has a particular, desirable mathematical structure (for details and the proof, see Schupbach 2017, Theorem 1; compare with Schupbach and Sprenger 2011, Theorems 1 and 2):[58]

$$\mathcal{E}_{SS}(e,h) = \frac{Pr(h|e) - Pr(h|\neg e)}{Pr(h|e) + Pr(h|\neg e)}.$$

Alternatively, **Positive Relevance** and **Irrelevant Conjunction** alone suffice to single out $\mathcal{E}_{SS}$ again as the best measure under a different set of structural constraints – including the rather strong condition that the resulting measure must be representable as a function purely of $Pr(h|e)$ and $Pr(h|\neg e)$ (Cohen 2015, Corollary 1; cf. Crupi and Tentori 2012, Theorem 1).[59]

Note that $\mathcal{E}_{SS}(e,h) > 0$ to the extent that $Pr(h|e) > Pr(h|\neg e)$.[60] That it is also thus the case that $\mathcal{E}_{SS}(e,h) > 0$ to the extent that $Pr(e|h) > Pr(e)$, in line with our explication of the basic Peircean sentiment, can readily be seen from the following formal result (where $\Leftrightarrow$ denotes inter-derivability):

$$\mathcal{E}_{SS}(e,h) = \frac{Pr(h|e) - Pr(h|\neg e)}{Pr(h|e) + Pr(h|\neg e)} > 0$$

$$\Leftrightarrow Pr(h|e) = \frac{Pr(h)Pr(e|h)}{Pr(e)} > \frac{Pr(h)Pr(\neg e|h)}{Pr(\neg e)} = Pr(h|\neg e)$$

$$\Leftrightarrow \frac{Pr(e|h)}{Pr(e)} > \frac{Pr(\neg e|h)}{Pr(\neg e)} = \frac{1 - Pr(e|h)}{1 - Pr(e)}$$

$$\Leftrightarrow Pr(e|h) - Pr(e|h)Pr(e) > Pr(e) - Pr(e|h)Pr(e)$$

$$\Leftrightarrow Pr(e|h) > Pr(e).$$

---

[58] Other Bayesian measures of power have been proposed. Good (1960) and McGrew (2003) independently offer defenses of a particular measure that we will reconsider in another context, in §3.4; and Crupi and Tentori (2012) defend their own candidate measure. For a comparison of these measures and their merits as explications of the Peircean notion of power, the reader is referred to (Schupbach and Sprenger, 2011; Schupbach, 2011a; Crupi and Tentori, 2012; Schupbach and Sprenger, 2014; Cohen, 2015, 2016; Sprenger and Hartmann, 2019).

[59] Sprenger and Hartmann (2019, Theorem 7.3) have recently put forward another compelling uniqueness theorem for $\mathcal{E}_{SS}$, which makes only minimal assumptions regarding the resulting measure's mathematical structure while relying on the following condition:

Independent Background Theories. If $Pr(h|e \wedge t) = Pr(h|e)$ and $Pr(h|\neg e \wedge t) = Pr(h|\neg e)$, then $\mathcal{E}(e,h) = \mathcal{E}(e,h|t)$.

Although **Independent Background Theories** arguably strays a bit further from the original Peircean sentiment, Sprenger and Hartmann (2019, p. 198) defend it by arguing that it explicates the following intuitive idea: "if a theory t is irrelevant to the interaction between explanans h and explanandum e (and its negation $\neg e$), then conditionalizing on t does not affect the degree of explanatory power."

[60] Indeed, $\mathcal{E}_{SS}$ is *ordinally equivalent* to the "posteriors ratio" measure $\mathcal{E}_{PR}(e,h) = Pr(h|e)/Pr(h|\neg e)$, meaning that $\mathcal{E}_{SS}$ and $\mathcal{E}_{PR}$ agree on all rankings: $\mathcal{E}_{SS}(\phi_1,\psi_1) \gtreqless \mathcal{E}_{SS}(\phi_2,\psi_2)$ if and only if $\mathcal{E}_{PR}(\phi_1,\psi_1) \gtreqless \mathcal{E}_{PR}(\phi_2,\psi_2)$, for all propositions $\phi_1, \phi_2, \psi_1, \psi_2$.

Along the same lines, one can demonstrate that $\mathcal{E}_{SS}(e, h) = 0$ if and only if $Pr(e|h) = Pr(e)$ and $\mathcal{E}_{SS}(e, h) < 0$ to the extent that $Pr(e|h) < Pr(e)$. And it is similarly straightforward to prove that $\mathcal{E}_{SS}(e, h)$'s maximal value of 1 is reached if and only if $Pr(e|h) = 1$ (per **Maximality**), whereas its minimal value of -1 is reached if and only if $Pr(e|h) = 0$.

With this measure of explanatory power in hand, we're able to investigate the normative implications of explanatory reasoning formally, from within the Bayesian perspective. To begin, we can continue the above demonstration and further explore the implications of the observation that some h has positive degree of power over explanandum e:

$$\mathcal{E}_{SS}(e, h) > 0 \Leftrightarrow Pr(e|h) > Pr(e)$$
$$\Leftrightarrow Pr(e|h) > Pr(e|h)Pr(h) + Pr(e|\neg h)Pr(\neg h)$$
$$\Leftrightarrow Pr(e|h) > Pr(e|\neg h) \tag{L}$$
$$\Leftrightarrow Pr(h|e) > Pr(h). \tag{C}$$

Note that $Pr(e|h)Pr(h) + Pr(e|\neg h)Pr(\neg h)$ constitutes a weighted average of $Pr(e|h)$ and $Pr(e|\neg h)$; this observation clarifies how (L) follows from its preceding line. (C) follows straightforwardly from $Pr(e|h) > Pr(e)$ by **Bayes's Theorem**.

(L) and (C) are both noteworthy normative implications.[61] To the extent that an hypothesis is able to provide a powerful explanation of the evidence in question, (C) reveals that the evidence confirms that hypothesis. This provides a clear sense in which the judgment that a hypothesis is positively explanatory of the evidence constitutes a reason to favor that hypothesis. (L) additionally clarifies that positive assessments of explanatory power correspond with likelihood comparisons. If h provides a positively powerful explanation of e, then e is more expected given the truth of h than given its falsity. Likelihood comparisons like (L) are thought to be foundational to the very notion of evidential support and favoring by some philosophers of science (Hacking, 1965; Sober, 2008, 2019) as well as by an entire "Likelihoodist" school of statistical thought (Birnbaum, 1962; Edwards, 1992; Royall, 1997). Let's now apply our Bayesian explication of explanatory power to the study of one particularly prevalent pattern of scientific reasoning.[62]

---

[61] See Brössel (2015, §3.1) for an alternative defense of the normative merits of our measure, building upon the Bayesian convergence theorem of Gaifman and Snir (1982).

[62] See Schupbach (2016) for an application of this notion of explanatory power to "explaining-away arguments" – another common pattern of reasoning. Glass (2016) provides an alternative Bayesian approach to explaining-away.

3.2.3 Inference to the Best (i.e., Most Powerful) Explanation

Scientists often evaluate theories by comparing their abilities to explain the evidence. Examples we've already cited fit this description. Geologists reason to the occurrence of an earthquake millions of years ago because this event would, more than any other candidate hypothesis, explain various deformations in layers of bedrock. Darwin's *Origin* provides an additional, particularly powerful example. Indeed, this work can be fairly characterized as one extended argument along these lines, in which Darwin repeatedly supports Natural Selection by arguing that it provides a superior explanation to a Design hypothesis with respect to a variety of biological explananda. The eminent historian and philosopher of science Ernan McMullin (1992) finds such a central role in scientific practice for this style of reasoning that he dubs it "the inference that makes science."

Philosophers label this basic style of reasoning "Inference to the Best Explanation" (IBE) (Harman, 1965; Lipton, 2004). When inferring a best explanation, an agent reasons to an explanatory hypothesis based upon the premise that it provides a better potential explanation of some given evidence than any other available, competing explanatory hypothesis. As such, IBE can be identified with the following inference form:

(P$_1$) Explanandum e obtains.

(P$_2$) Hypotheses $h_1, h_2, \ldots,$ and h_n offer competing explanations of e.

(P$_3$) Of these, h_i provides the best explanation of e.

∴ (C) h_i.

Clarifying the pattern of IBE in this way still leaves this form of reasoning ambiguous in at least one important sense. Premise (P$_3$) refers to a difference in explanatoriness – or explanatory goodness – between considered hypotheses. But explanatoriness is famously evaluated along different dimensions, corresponding to the various acclaimed explanatory virtues. Potential explanations may be prized for their great simplicity, unification, generality, power, coherence, consilience, or some combination of these (or other) virtues. One immediate consequence of this, often overlooked by IBE's commentators, is that the nature and value of an inference to the best explanation will depend upon the notion of explanatoriness at work therein. As a general category, IBE is polymorphous. There are at least as many distinct forms of IBE as there are distinct dimensions along which hypotheses may be judged more explanatory than one another. Inference to the Best (i.e., Most Unifying) Explanation, for example, differs (prima facie, quite substantially) from Inference to the Best (i.e., Simplest) Explanation.

Any careful articulation and evaluation of IBE must accordingly specify the notion of explanatoriness at work in (P_3). Here, we'll focus on a version, IBE_p, that evaluates explanatoriness as power à la Peirce. With measure $\mathcal{E}_{SS}$ in hand, IBE_p can be explicated as follows:

> Explanandum e obtains.
> Hypotheses $h_1, h_2, \ldots,$ and h_n offer competing explanations of e.
> $\mathcal{E}_{SS}(e, h_i) > \mathcal{E}_{SS}(e, h_j)$, where $1 \leq i, j \leq n$ and for all $j \neq i$.
> $\therefore\ h_i$.

Note that IBE_p, as we've stated it here, does not apply when there is no single most powerful explanation for e among the candidates – e.g., when there is a tie for the most powerful explanation.

To evaluate IBE_p, we first can investigate the probabilistic implications of the central premise that h_i has more power than any other candidate hypothesis (note that the second line below follows from the ordinal equivalence of $\mathcal{E}_{SS}$ and $\mathcal{E}_{PR}$ as described in footnote 60):

$$
\begin{aligned}
\mathcal{E}_{SS}(e, h_i) &= \frac{Pr(h_i|e) - Pr(h_i|\neg e)}{Pr(h_i|e) + Pr(h_i|\neg e)} > \frac{Pr(h_j|e) - Pr(h_j|\neg e)}{Pr(h_j|e) + Pr(h_j|\neg e)} \\
&= \mathcal{E}_{SS}(e, h_j) \\
&\Leftrightarrow \frac{Pr(h_i|e)}{Pr(h_i|\neg e)} > \frac{Pr(h_j|e)}{Pr(h_j|\neg e)} \\
&\Leftrightarrow \frac{Pr(e|h_i)Pr(\neg e)}{Pr(\neg e|h_i)Pr(e)} > \frac{Pr(e|h_j)Pr(\neg e)}{Pr(\neg e|h_j)Pr(e)} \\
&\Leftrightarrow Pr(e|h_i) - Pr(e|h_i)Pr(e|h_j) > Pr(e|h_j) \\
&\qquad - Pr(e|h_i)Pr(e|h_j) \\
&\Leftrightarrow Pr(e|h_i) > Pr(e|h_j).
\end{aligned}
$$

The upshot is that the most powerful explanatory hypothesis h_i will have the highest likelihood. This result clarifies a sense in which reason does indeed favor the h_i over its competitors, since this hypothesis's likelihood $Pr(e|h_i)$, as we have seen, positively relates to its overall probability in light of the evidence $Pr(h_i|e)$. All else (viz., prior probability) being equal, the greater a hypothesis's likelihood, the more probable it is in the light of explanandum e (McGrew, 2003, p. 558).

For a more detailed assessment of how well IBE_p performs, we can use computers to simulate this style of reasoning under certain specified conditions and gauge its "success rate" (Schupbach, 2017, §2.2). The scenarios that we wish to simulate will be ones in which reasoners are indeed inclined to apply

IBE$_p$ – that is, scenarios in which they are inclined to evaluate explanatory goodness purely in terms of power. These will, for example, be contexts in which other explanatory virtues like simplicity and scope don't clearly favor any of the candidate explanations – and so they don't enter substantially into the evaluation of explanatory goodness. With this in mind, the general procedure that the simulations execute is as follows:[63]

1. Assign priors $Pr(h_i)$ and likelihoods $Pr(e|h_i)$ to each of n mutually exclusive explanatory hypotheses. Likelihoods are sampled randomly from the standard uniform distribution over $[0, 1]$, while priors are drawn randomly from normal distribution $\mathcal{N}(.5, .15^2)$ and then renormalized to sum to one.
2. Assign objective chances to hypotheses. $Ch(h_i) = Pr(h_i) + X$, where $X \sim \mathcal{N}(0, \sigma^2)$ (again, renormalized to ensure that they sum to 1).
3. Using the respective values of $Ch(h_i)$, randomly (stochastically) select the "true" hypothesis h_j from $\{h_1, h_2, \ldots, h_n\}$.
4. Using the value of $Pr(e|h_j)$ (the likelihood associated with the true hypothesis), check whether e "occurs." If e occurs, continue with steps 5–7; otherwise, end this iteration.
5. Check which of the n hypotheses has the greatest explanatory power; that is, find h_k where $\mathcal{E}_{SS}(e, h_k) > \mathcal{E}_{SS}(e, h_i)$ for all $i \neq k$.
6. Check which of the n hypotheses is the most probable in light of e; that is, find h_l where $Pr(h_l)Pr(e|h_l) > Pr(h_i)Pr(e|h_i)$ for all $i \neq l$.
7. If $h_k = h_j$, count this as a case where the most explanatorily powerful hypothesis matches the true hypothesis; if $h_l = h_j$, count this as a case where the most probable hypothesis matches the true hypothesis.

Steps 1–7 constitute one iteration of the simulation. In any run of the simulation, a value of n (ranging from 2 to 10) is fixed across all iterations. To secure accurate results, steps 1–7 are iterated as many times as it takes for there to be 1,000,000 iterations in which e occurs (see step 4). The computer then provides estimates of how often the hypothesis with the greatest power is true and how often the hypothesis with the greatest posterior probability is true. Such success rates are calculated as the number of times that one gets such a match divided by the total number of instances in which e occurs (i.e., 1,000,000).

In step 1, priors are drawn randomly from normal distribution $\mathcal{N}(\mu = .5, \sigma = .15)$ so that they tend to cluster in value. Sampling priors in this way models the idea discussed earlier that explanatory virtues other than power (e.g., simplicity) are not clearly favoring some candidate hypotheses over the

---

[63] These simulations are based closely upon those devised and reported by Glass (2012).

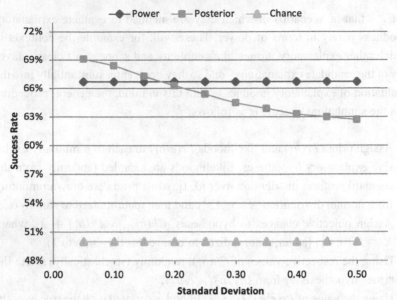

Figure 8 Success rate of IBE_p in contexts with no catchall, compared to those of the posterior probability and chance guessing ($n = 2$).

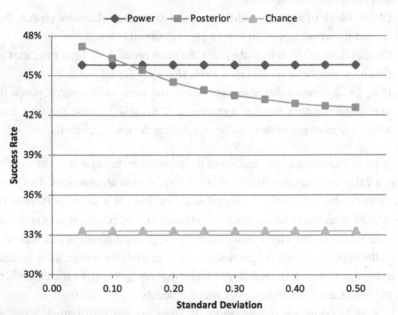

Figure 9 Success rate of IBE_p in contexts with a catchall, compared to those of the posterior probability and chance guessing ($n = 2$).

others. Regarding step 2, the standard deviation σ for the error term X is allowed to vary between simulation runs, allowing us to experiment with the average inaccuracy of an agent's priors. Finally, steps 1–7 model a situation in which the reasoning agent is working with a partition, a situation in which the reasoner knows with certainty that exactly one of the hypotheses being considered is true. Of course, this is not always a realistic assumption; often, we require a "catchall" hypothesis standing in for all of the various alternative hypotheses unimagined by (and so unavailable to) the reasoner. Accordingly, simulations are run both for situations that do and situations that don't involve a catchall hypothesis.[64]

Results in the case where $n = 2$ are shown in Figure 8 (contexts with no catchall) and Figure 9 (contexts including a catchall). These figures display the respective success rates of IBE_p and posterior probabilities across the same range of settings for average error (σ) in prior probabilities.

What do we learn about the normative implications of IBE_p from these results? Both figures reveal that IBE_p does markedly better than chance at hitting upon the truth.[65] In this sense, IBE_p's premises truly offer inductive support (albeit, defeasible support) for the truth of a corresponding conclusion. In some select cases (e.g., Figure 8), IBE_p's conclusions are even more often true than false; however, such a result need not hold in general to warrant this pattern of inference. By reasoning via IBE_p we get a leg up on discovering true hypotheses, and this holds even in cases where that leg up still doesn't single out hypotheses that are more often true than false.

This may seem like a disappointing result until one remembers that IBE_p itself doesn't recommend any specific epistemic stance toward its conclusion. In particular, there is no reason to think that the conclusion of an IBE_p is always to be believed outright. We seen that Peirce himself described his explanatory inference as merely providing "reason to suspect that h is true." Our computer simulations suggest one way to go beyond Peirce and provide more precise advice regarding the stance one should take toward IBE_p's conclusions. On the foundations of Ramsey's "logic of truth" and calibration principles (§2.3), IBE_p's success rates can be used to calibrate the degrees of confidence

---

[64] When a catchall hypothesis is included in a simulation run, its prior probability may stray from the priors of the considered hypotheses. The latter are sampled randomly from normal distribution $\mathcal{N}(\mu = .5, \sigma = .15)$, but the catchall's prior is drawn randomly from a uniform distribution over $[0, 1]$ (all values are then renormalized to sum to 1). Any included catchall hypothesis can be chosen as the true hypothesis h_j in step 3, but it cannot be chosen as the most powerful (h_k in step 5) or the most probable (h_l in step 6) of the hypotheses. Appropriately, for any iteration in which the catchall is true, IBE_p and reasoning to the most probable hypothesis are both sure to miss the truth.

[65] This is true whether we include a catchall or not, no matter how inaccurate priors are allowed to become (i.e., no matter how large the σ), and across the other values of n.

agents have in conclusions reached by that inference form. For example, referring to Figures 8 and 9, if an agent is reasoning to one of two hypotheses purely on the basis of IBE_p, then calibrating to the success rates of such inferences would lead that agent to have confidence $\approx .67$ or $\approx .46$ in the most explanatorily powerful hypothesis, depending on whether the context calls for a catchall or not.

Finally, note that the success rate of IBE_p can easily exceed that of reasoning to the most probable hypothesis depending on the inaccuracy of one's prior probabilities. The "crisscrossing" displayed in both figures demonstrates this latter qualitative result.[66] Interestingly then, this Bayesian account of explanatory inference suggests that we sometimes do better when it comes to finding the truth by following likelihood-tracking judgments of explanatory power to the neglect of potentially inaccurate prior probabilities.

3.3 Robustness Analysis

Scientists often need to confirm that certain results they are detecting are not simply artifacts of the particular means used to detect them. In such cases, they may attempt to duplicate those results using diverse means. To the extent that they're successful in doing so, their results are shown to be *robust*. Robustness analysis [RA] is a mode of reasoning in which one supports an hypothesis via an analysis of the conditions under which a result proves robust.

Examples of RA from scientific practice abound. A well-known, seminal discussion of RA's usefulness in science comes from biologist Richard Levins (1966, p. 423). Levins specifically advocates the use of RA in biological modeling as a general means for deciphering, when using simplified models to study complex systems, whether a result "depends on the essentials of a model or on the details of the simplifying assumptions":

> [W]e attempt to treat the same problem with several alternative models each
> with different simplifications but with a common biological assumption.
> Then, if these models, despite their different assumptions, lead to similar
> results we have what we can call a robust theorem which is relatively free of
> the details of the model. Hence our truth is the intersection of independent
> lies.

---

[66] Again, this result holds robustly across various parameter settings, including whether to include a catchall and across the other values of n. Interestingly, it also turns out that the qualitative results of these simulations rely in no way on the means of sampling priors; even if we sample priors from the standard uniform distribution (and then renormalize so that they sum to one), we still get crisscrossing plots of percentage accuracies with increasingly inaccurate priors.

In the history of physics, the case of Brownian motion provides a particularly vivid example of RA. When suspended in a fluid medium, sufficiently small particles display continuous and random motion. Upon discovering this phenomenon, the botanist Robert Brown surmised that the movements were due to the particular (uniquely shaped) pollen granules he was observing. However, he found that the movements persisted across experiments using other particles – first with other types of pollen, then other organic materials, and eventually using inorganic particles (Brown, 1828). Over the next 75 years, other experimenters showed that such "Brownian motion" was robust over changes in the fluid medium, container used, means of suspending the particles, environmental conditions around the container, and so on. Eventually, scientists working in the wake of Einstein's *annus mirabilis* appealed to the robustness of Brownian motion in order to confirm that there were deeper, unobservable agitations of molecules within the medium – Perrin (1913, p. 83) likens this to the evident rocking of a far-off ship indicating imperceptibly distant waves on the sea.

In this case, the Brownian motion is the robust result, and the diverse means of detection are the various experiments used. The Brownian motion is notably robust across certain changes to the experimental apparatus (e.g., type of particle, medium, container, lighting) and sensitive to others (e.g., size of particle, temperature of medium). And it is upon analyzing these conditions of robustness that Perrin (1913, p. 86) says we are "forced to conclude," consonant with Einstein's molecular explanation, that there are internal, unobservable movements in the medium.

Other example RAs include cases from cognitive psychology (Stolarz-Fantino et al., 2003; Crupi et al., 2008), physics (Hacking, 1983; Cartwright, 1991; Mayo, 1996), biology (Culp, 1994; Plutynski, 2006; Weisberg and Reisman, 2008), climate science (Lloyd, 2010; Parker, 2011; Winsberg, 2018), epidemiology (Stegenga, 2009), and modeling in economics (Woodward, 2006; Kuorikoski et al., 2010). As the range of these cases makes evident, we're using the terms "results" and "means of detection" quite liberally. The results in question could be observations, measurements, predictions, theorems, and so on. Correspondingly, the means of detecting such results could include experiments, laboratory instruments, sensory modalities, derivations (from axioms, models, theories, etc.), axiomatic systems, computer simulations, and formal models among other things.

Some RAs are more empirically driven, analyzing the robustness of experimental and observational results. Others are more analytically or model driven, focusing on robust conclusions and simulation results. While philosophers of science have found it useful to distinguish particular varieties of RA (Woodward, 2006; Calcott, 2011), allowing them to examine peculiar features of each type, here we consider an account (first developed in Schup-bach 2018) that claims to unify most, if not all, cases of RA. Following Wimsatt (2011, p. 299), the aim is "to show the common features of these diverse methods." After developing and articulating the account informally, we'll provide a Bayesian explication and evaluation of RA so conceived.

3.3.1 RA-Diversity and Explanatory Discrimination

The intuition motivating the use of RA is that we may gain confirmation through *diversity*; certain hypotheses (e.g., Einstein's molecular explanation of Brownian motion) are supported to the extent that a result proves robust, and results prove robust to the extent that we detect them in diverse ways. To understand and evaluate this form of scientific reasoning, it's thus of the utmost importance to determine what precise sense of diversity is at work in RA.

Bayesians offer a variety of accounts of evidential diversity (Fitelson, 2001a; Bovens and Hartmann, 2003; McGrew, 2016). With few exceptions (Horwich, 1982; Howson and Urbach, 2006), these explicate diversity using some prob-abilistically precise notion of *independence*. Many of these accounts plausibly capture senses in which we speak of evidence as being diverse. Moreover, sev-eral of these accounts imply interesting normative results according to which diverse bodies of evidence may be specially confirmatory. Philosophers work-ing on RA have thus understandably drawn upon these independence-based accounts in attempting to articulate the sense of diversity at work in RA, or "RA-diversity" (Wimsatt, 1981, 1994; Lloyd, 2009, 2010; Kuorikoski et al., 2010; Stegenga and Menon, 2017).

Unfortunately, such independence-based notions of evidential diversity fun-damentally mismatch the notion of RA-diversity (Schupbach, 2015). That is, while these accounts describe important and fruitful senses in which evidence can be diverse, they fail to capture the concept at work in paradigmatic cases of RA from science. The general issue, in short, is that perfectly RA-diverse means of detection (across a variety of typical and canonical examples) can overall be quite similar to one another – think, for example, of experiments detecting Brownian motion under very slightly different conditions. Such means of detec-tion can be *strongly* dependent on one another and thus decidedly *not* diverse according to any of the alternative independence-based accounts.

Horwich (1982, pp. 118–122) proposes a very different account of evidential diversity. While this account is Bayesian and thus probabilistic, it doesn't make use of probabilistic independence. The central notion in Horwich's account is rather that of *elimination*; diverse bodies of evidence, according to Horwich, "tend to eliminate from consideration many of the initially most plausible, competing hypotheses" (p. 118). Probabilistically, Horwich represents "initially plausible, competing hypotheses" as those with substantial prior probabilities, and he identifies diverse evidence with evidence having low likelihoods conditional on competing hypotheses (since it is stipulated as eliminating these hypotheses).

The normative result that more diverse bodies of evidence better confirm a hypothesis, on Horwich's account, may then be demonstrated as follows. Let e_D describe a more diverse body of evidence than e_N relative to target hypothesis h_1 and its competitors $h_2, h_3, \ldots, h_k$. Horwich stipulates that these hypotheses form a partition. For simplicity, Horwich also assumes that h_1 implies e_D and e_N, so that we have $Pr(e_D|h_1) = Pr(e_N|h_1) = 1$. Then we may compare how well each body of evidence confirms h_1 by comparing $Pr(h_1|e_D)$ to $Pr(h_1|e_N)$ in the following way:

$$
\frac{Pr(h_1|e_D)}{Pr(h_1|e_N)} = \frac{Pr(e_D|h_1)}{Pr(e_N|h_1)} \times \frac{Pr(e_N)}{Pr(e_D)} = \frac{Pr(e_N)}{Pr(e_D)}
$$
$$
= \frac{Pr(h_1) + Pr(e_N|h_2)Pr(h_2) + \ldots + Pr(e_N|h_k)Pr(h_k)}{Pr(h_1) + Pr(e_D|h_2)Pr(h_2) + \ldots + Pr(e_D|h_k)Pr(h_k)}.
$$

Comparing like terms between the numerator and denominator of this ratio, the only terms that differ are the likelihoods relative to h_1's competitors. Consequently, to the extent that all of those hypotheses with considerable values of $Pr(h_i)$ are such that $Pr(e_D|h_i) < Pr(e_N|h_i)$, it will tend to be the case that $Pr(h_1|e_D) > Pr(h_1|e_N)$. But that's just to say that the more diverse the evidence in this case, the more confirmation it will tend to bestow upon the target hypothesis h_1.

Horwich's account has its shortcomings. For one thing, we require a subtler demonstration of the normative impact of diversity, one that doesn't make the problematic assumptions that h_1 implies e_D and e_N, and that h_1 and its competitors form a partition. These assumptions rarely if ever all hold in standard cases of RA (they certainly don't hold in the Brownian motion case). Additionally, as Fitelson (1996) suggests, Horwich's discussion doesn't offer an account of evidential diversity so much as an explication of the logical effects of diversity. A more satisfying theory would first do the conceptual groundwork of explicating the notion of diversity at work, and only then offer a more informed exploration of the logical effects of diversity in this sense. This is what we'll pursue presently.

Despite the above shortcomings, Horwich's basic association of diversity with elimination does align nicely with the role of diversity in RA. For example, each new, diverse means of detecting Brownian motion cited by Perrin has the effect of *eliminating* a potential explanation of this phenomenon that would confound the case for the molecular explanation (the target hypothesis) H. Had Perrin rested his case after reporting only that Brown observed various pollen granules in motion when suspended in water, too many competing explanations of the observed motion would be left standing, blocking the case for H. One is hardly compelled to infer H from this single observation; there are too many alternative hypotheses remaining at this stage for H (or any of H's competitors) to be well supported by the result. So, instead of stopping there, Perrin mentions experiment after experiment detecting the result, these differing from one another only subtly, but insofar as each is able to eliminate different sets of H's competitors.

Similarly, in model-based cases of RA, a single model or simulation exemplifying behavior in accordance with some theorized principle[67] could hardly be said to lend strong support for this general principle. There are typically far too many ways that any such formal model diverges from the real world scenario(s) it's trying to represent. And the salient behavior of the model could be driven by any subset of such idealizing and simplifying assumptions. To eliminate such competing explanations of the model's behavior, a variety of other models are thus employed, which likewise attempt to model behavior in accordance with the theorized principle while differing from one another only subtly, but insofar as they toggle the simplifying and idealizing assumptions involved.

In such standard cases of RA, means of detecting the robust result can be similar in nearly all respects. What matters is that they differ in some *relevant* respect. Specifically, each incremental step of a RA cites an additional means of detection that has an eliminative power, the power to discriminate between the target explanation h of the robust result r and some competing potential explanation(s) h'. Such means of detection are *explanatorily discriminating between h and h'* in the sense that h would explain well our detecting r via this new means, whereas h' would explain well our *failing to detect r* by this means. More generally, where r_k denotes the proposition that r is detected using the kth means of detection:

RA-diversity. n means of detecting r are all RA-diverse with respect to h and its competitors to the extent that $r_1, r_2, ..., r_n$ can be put into a sequence for which any member is explanatorily discriminating between h and some competing explanation(s) not yet ruled out by the prior members of that sequence.

This account of RA-diversity fits well with standard cases of RA in science. Perrin's hypothesis that Brownian motion is due to internal, invisible

---

[67] E.g., see Weisberg and Reisman's (2008) discussion of RA and the Volterra Principle.

movements in the medium would explain well our observations of such motion using inorganic material, whereas the vital force hypothesis would explain well our failing to observe the result in this case; thus, this experimental detection explanatorily discriminates between these hypotheses. That a simulation demonstrates behavior in accordance with a theorized principle without involving an unrealistic assumption (e.g., that prey cannot take cover) is explained well by that principle itself (in conjunction with the hypothesis that the model is accurately depicting the real-world behavior), but the competing explanation that this behavior is attributable to the unrealistic assumption in question would rather provide a strong explanation of our *failing* to observe the behavior using such a model; such a model thus explanatorily discriminates between these potential explanations.

In sum, RA-diverse means of detection provide sequences of explanatorily discriminating bits of evidence, which successively eliminate more and more of h's competitors. As desired, on this account, it is really not so relevant whether means of detection are *strongly* diverse or sufficiently heterogeneous in some absolute sense, apart from considered hypotheses. What matters for RA-diversity is that the means (which may actually be quite similar in most respects) are different in just the sense required to rule out h's salient competitors.

3.3.2 Interlude: Hypothesis Competition and Defeat

A successful increment of RA is one that's able to discriminate between the target hypothesis h and at least one of $h's$ competitors. But what exactly does it take for hypotheses to compete?[68] This interlude briefly explores this question and thereby offers an additional example of Bayesianism applied to the epistemology of science. We'll then circle back to our discussion of RA with a more precise, accurate, and useful explication of hypothesis competition in hand.

The concept that we're interested in clarifying here is *competition in an epistemic sense*; in RA, we consider hypotheses to be competitors insofar as reason compels us to accept at most one of these.[69] Under what conditions is this the case? A prima facie obvious answer is that we rationally ought to choose between hypotheses when they are mutually exclusive. However, as discussed in Schupbach and Glass (2017), it's far too easy to think of counterexamples to

---

[68] The research presented here was carried out as a collaboration between the author and David H. Glass.

[69] Hypotheses may also compete in objective, ontic senses (e.g., if it's metaphysically impossible for them simultaneously to be true). However, we're interested in what it means for hypotheses to compete in some agent's epistemic economy, and such epistemically competing hypotheses may or may not compete in a deeper, ontic sense.

this idea. Hypotheses may be consistent with one another and yet still compete epistemically. Notably, this is true of the various hypotheses in our examples of RA (e.g., multiple of the potential explanations of Brownian motion could simultaneously be true).

In fact, there are at least two distinct ways in which hypotheses may epistemically compete, even if they are consistent with one another:

1. Hypotheses might compete *directly* by virtue of disconfirming one another, whether they do so decisively or not.
2. Hypotheses may compete *indirectly*, via some specified evidence, when they both relate to that evidence in such a way that accepting either hypothesis neutralizes the evidential support for the other, or even turns that evidence against the other.

These two "pathways" via which hypotheses may compete correspond to the distinct notions of *rebutting* and *undercutting* defeat (Pollock, 1987). Moreover, as we'll presently discover, the Bayesian approach allows us to explicate competition along each pathway as a matter of degree.

Let's begin by explicating what it takes for hypotheses to *all-out* compete by virtue of *decisively* defeating one another in either sense. First, hypotheses h_1 and h_2 may be said to all-out compete along the direct pathway when they decisively rebut one another. This extreme condition corresponds to the case in which the hypotheses are perceived to mutually exclude one another. Probabilistically:

$$Pr(h_1 \wedge h_2) = 0. \tag{DR}$$

Second, h_1 and h_2 may be said to all-out compete indirectly via evidence e when each decisively undercuts the other's evidential support (from e). Complete undercutting may be interpreted as the case in which evidence e that supports h_1 ends up being irrelevant to h_1 conditional on the undercutting defeater h_2; formally: $Pr(h_1|e) > Pr(h_1)$, but $Pr(h_1|e \wedge h_2) = Pr(h_1|h_2)$. However, there is a much more extreme version yet of this scenario: when supposing h_2 not only negates the support e provides for h_1, but turns e decisively against h_1:

$$Pr(h_1|e) > Pr(h_1), \text{ but } Pr(h_1|e \wedge h_2) = 0 < Pr(h_1|h_2). \tag{DU}$$

(This final inequality follows if we stipulate that this is not also a case of decisive rebutting defeat.)

Our explications of decisive rebutting (DR) and decisive undercutting (DU) can be combined into a very simple and plausible Bayesian explication of decisive defeat (DD): h_1 and h_2 decisively defeat one another with respect to e if and only if $Pr(h_1 \wedge h_2|e) = 0$. That this condition appropriately incorporates both (DR) and (DU) can be seen by expanding $Pr(h_1 \wedge h_2 | e)$ via **Bayes's Theorem**:

$$Pr(h_1 \wedge h_2 | e) = \frac{Pr(h_1 \wedge h_2) Pr(e | h_1 \wedge h_2)}{Pr(e)} = 0. \tag{DD}$$

(DR) and (DU) describe the two ways that $Pr(h_1 \wedge h_2 | e)$ can equal zero – each condition corresponding to one of the two factors in the numerator equaling zero.[70]

Developing this idea into an account of the *degree to which h_2 defeats h_1* with respect to e requires explications of the degree to which h_2 rebuts h_1 and the degree to which h_2 reduces the support e otherwise provides for h_1. Plausible accounts along these lines may be developed using a measure of incremental confirmation (see §3.1) – that is, some way of measuring, for any h and e, the strength of the inequality $Pr(h|e) > Pr(h)$. There are many importantly distinct measures available (Fitelson, 1999; Crupi et al., 2007). Here, we'll make use of the popular and well-regarded "log-likelihood" measure. This measure calculates the degree to which e supports h as follows:[71]

$$C_l(e, h) = \log \left[\frac{Pr(e|h)}{Pr(e|\neg h)} \right].$$

Using C_l, we explicate the degree to which an hypothesis h_2 directly rebuts another h_1 as the degree to which h_2 confirms that h_1 is false:

$$\mathcal{D}_{REB}(h_2, h_1) = C_l(h_2, \neg h_1) = \log \left[\frac{Pr(h_2 | \neg h_1)}{Pr(h_2 | h_1)} \right].$$

As desired, $\mathcal{D}_{REB}(h_2, h_1)$ takes its extreme value in the case of decisive rebutting (DR). This follows from the fact that it takes its maximal value (∞; see footnote 71) exactly when $Pr(h_2 | h_1)$ – and thus $Pr(h_1 \wedge h_2)$ – equals 0. Since hypotheses compete along the direct pathway to the extent that they rebut one another, the degree of direct competition between h_1 and h_2 can then be measured as the average degree to which these hypotheses rebut one another:

$$\text{Comp}_D(h_1, h_2) = (\mathcal{D}_{REB}(h_1, h_2) + \mathcal{D}_{REB}(h_2, h_1))/2.$$

To measure the degree to which h_2 reduces the support e otherwise provides for h_1, we can use the same confirmation measure and compare two factors: the degree to which e supports h_1 apart from considerations of h_2, $C_l(e, h_1)$,

---

[70] (DU) requires that $Pr(h_1 | e \wedge h_2) = 0$ and $Pr(h_1 | h_2) \neq 0$. But then, given (DU), $Pr(h_1 | e \wedge h_2) = Pr(h_1 | h_2) Pr(e | h_1 \wedge h_2)/Pr(e | h_2) = 0$ implies $Pr(e | h_1 \wedge h_2) = 0$.

[71] We can ensure that C_l is defined over the extended real numbers, $\mathbb{R} \cup \{-\infty, +\infty\}$, by defining $C_l(e, h) = \infty$ if $P(e | \neg h) = 0$ (in which case, $P(e) \neq 0$ implies that $P(e | h) \neq 0$), and $C_l(e, h) = -\infty$ if $P(e | h) = 0$ (in which case, $P(e) \neq 0$ implies that $P(e | \neg h) \neq 0$).

and the degree to which e supports h_1 conditional on h_2, $C_l(e, h_1|h_2)$. The difference between these terms measures the degree to which h_2 undercuts the support from e to h_1:

$$\mathcal{D}_{UND}(h_2, h_1/e) = C_l(e, h_1) - C_l(e, h_1|h_2)$$

$$= \log\left[\frac{Pr(e|h_1)}{Pr(e|\neg h_1)}\right] - \log\left[\frac{Pr(e|h_1 \wedge h_2)}{Pr(e|\neg h_1 \wedge h_2)}\right].$$

As desired, $\mathcal{D}_{UND}(h_2, h_1/e)$ takes its extreme value in the case of decisive undercutting (DU). This can be seen from the fact that it approaches its maximal value (∞) as $Pr(e|h_1 \wedge h_2)$ goes to 0 (see footnote 70). Since hypotheses compete indirectly via evidence to the extent that they undercut one another's evidential support, the degree of indirect competition between h_1 and h_2 apropos e can be measured as average degree of undercutting defeat:

$$\text{Comp}_I(h_1, h_2/e) = (\mathcal{D}_{UND}(h_1, h_2/e) + \mathcal{D}_{UND}(h_2, h_1/e))/2.$$

To derive a combined measure of the *net* degree to which h_2 defeats h_1 with respect to e, we may simply add these two components together:[72]

$$\mathcal{D}(h_2, h_1/e) = \mathcal{D}_{UND}(h_2, h_1/e) + \mathcal{D}_{REB}(h_2, h_1)$$

$$= \left(\log\left[\frac{Pr(e|h_1)}{Pr(e|\neg h_1)}\right] - \log\left[\frac{Pr(e|h_1 \wedge h_2)}{Pr(e|\neg h_1 \wedge h_2)}\right]\right)$$

$$+ \log\left[\frac{Pr(h_2|\neg h_1)}{Pr(h_2|h_1)}\right].$$

Just as we were able to combine the two types of decisive defeat into a single probabilistic criterion above, the following theorem shows that net degree of defeat reduces to a single, familiar criterion:[73]

Theorem 1. $\mathcal{D}(h_2, h_1/e) = C_l(h_2, \neg h_1|e)$.

In words, the degree to which h_2 defeats h_1 with respect to e just amounts to the degree to which h_2 disconfirms h_1 (i.e., confirms $\neg h_1$) given e.

The overall degree to which two hypotheses h_1 and h_2 compete in light of some evidence e (taking into account both the direct and indirect pathways to epistemic competition) is determined as the average net degree to which h_1 and h_2 *defeat* one another in the light of e. This idea combined with

---

72 Note that $\mathcal{D}_{UND}$ will not be defined if either $Pr(h_1 \wedge h_2) = 0$ or $Pr(\neg h_1 \wedge h_2) = 0$. This unfortunate result will be rectified in the measure of degree of competition.

73 The proof can be easily extrapolated from Schupbach and Glass's (2017, pp. 815–816) Theorem 1 and its corresponding proof.

Theorem 1 allows us then to explicate competition in Bayesian confirmation-theoretic terms:

$$\text{Comp}(h_1, h_2/e) = \frac{C_l(h_1, \neg h_2|e) + C_l(h_2, \neg h_1|e)}{2}.$$

Explicating degree of competition purely in terms of C_l (rather than $\mathcal{D}$) has the benefit of resulting in a measure that's appropriately defined in extreme cases where $Pr(h_1 \wedge h_2) = 0$ or $Pr(\neg h_1 \wedge h_2) = 0$.

The upshot is that hypotheses compete epistemically with one another relative to some evidence to the extent that they disconfirm one another in light of that evidence. Degree of competition is appropriately symmetric; the degree to which h_1 and h_2 compete in light of e is necessarily equal to the degree to which h_2 and h_1 compete in light of e: $\text{Comp}(h_1, h_2/e) = \text{Comp}(h_2, h_1/e)$. Qualitatively, hypotheses h_1 and h_2 compete (to some degree) relative to e if and only if $\text{Comp}(h_1, h_2/e) > 0$, and thus exactly when h_1 and h_2 disconfirm each other (to some degree) conditional on e.[74] Note that the measure $\text{Comp}(h_1, h_2/e)$ appropriately incorporates decisive defeat (DD), $Pr(h_1 \wedge h_2|e) = 0$, as the limiting case in which competition is maximal.

This account of hypothesis competition has some illuminating implications. Of course, as desired, hypotheses can compete – and even all-out compete – even when they are not mutually exclusive. An important, related point is that competition is relative to evidence; two hypotheses can compete with respect to one set of evidence while not competing with respect to another. This account also implies the plausible result that two hypotheses cannot possibly compete with respect to any evidence if one entails the other.[75]

3.3.3 A Bayesian Evaluation of Robustness Analysis

Returning now to the main topic, a successful *increment* of RA can be described via the following conditions:

Past Detections. We have a result r that we have detected using $n - 1$ different means. Let e denote the conjunction $r_1 \wedge r_2 \wedge \ldots \wedge r_{n-1}$.

Success. Hypothesis h (the hypothesis that we seek to confirm by our RA) explains this coincidence, but so does another hypothesis (or class of hypotheses) h'.

Competition. h and h' compete with one another, with respect to e.

---

[74] Similarly, $\text{Comp}(h_1, h_2/e) = 0$ exactly when h_1 and h_2 are confirmationally irrelevant given e, and $\text{Comp}(h_1, h_2/e) < 0$ iff h_1 and h_2 *confirm* each other conditional on e.

[75] Proof. Let $h_1 \vDash h_2$. Then, for arbitrary e (logically independent of h_1 and h_2), $Pr(h_2|h_1 \wedge e) = 1$ and so $Pr(h_2|h_1 \wedge e) \geq Pr(h_2|\neg h_1 \wedge e)$. This in turn means that $C_l(h_2, \neg h_1|e) \leq 0$, from which it also follows that $C_l(h_1, \neg h_2|e) \leq 0$. Consequently, $\text{Comp}(h_1, h_2/e) \leq 0$.

RA-Diversity. There is an nth means of potentially detecting r, which *explanatorily discriminates* between h and h'. That is, h would strongly explain r_n, and h' would strongly explain $\neg r_n$.

Robustness. The new means of detection concurs with prior means; that is, r_n.

The foregoing §3.3.2 developed a precise, confirmation-theoretic account that allows us to articulate the above **Competition** condition precisely. However, in order to explore the normative implications of RA from a Bayesian perspective, we still need more precise explications of **Success** and **RA-Diversity**. These conditions explicitly involve explanatory considerations. To clarify the epistemic implications of these considerations, we make use of our preferred account of explanatory power (§3.2.2):

$$\mathcal{E}_{SS}(e, h) = \frac{Pr(h|e) - Pr(h|\neg e)}{Pr(h|e) + Pr(h|\neg e)}.$$

Specifically, we formalize the judgment that some h (potentially) explains some e as a positive degree of explanatory power, $\mathcal{E}(e, h) > 0$, and we formalize the judgment that some h strongly (potentially) explains some e as near-maximal degree of explanatory power, $\mathcal{E}(e, h) \approx 1$.

These explicative choices, combined with our Bayesian accounts of explanatory power and epistemic competition, result in the following formal renderings of the conditions for a successful increment of RA:

Past Detections. We are given e (shorthand for $r_1 \wedge r_2 \wedge \ldots \wedge r_{n-1}$).

Success. $\mathcal{E}(e, h) > 0$ and $\mathcal{E}(e, h') > 0$.

Competition. $\text{Comp}(h, h'/e) > 0$.

RA-Diversity. $\mathcal{E}(r_n, h|e) \approx 1$, $\mathcal{E}(\neg r_n, h'|e) \approx 1$.

Robustness. We learn that r_n.

We're now in a position to investigate the normative impact of a successful increment of RA from the Bayesian perspective.

Mutually Exclusive Competitors

While the recognition of mutual exclusivity between hypotheses is not required for them to compete, it is sufficient (§3.3.2). And it proves informative to provide cases of mutual exclusivity with their own normative analysis. After doing so, we will examine the import of RA under the more general condition for hypothesis competition (i.e., **Competition** above).

Assume then that h and h' compete in the sense of being recognized as mutually exclusive. In this case, we can partition the space of possibilities into the set $\{h, h', c\}$, where c is a catchall hypothesis – logically equivalent to $\neg(h \vee h')$. To gauge the epistemic import of an increment of RA, we compare $Pr(h|e \wedge r_n)$ and

$Pr(h|e)$ in order to clarify the conditions under which the former term might be greater than the latter, and what determines the extent of this inequality. We may use the above partition to expand $Pr(h|e \wedge r_n)$ and $Pr(h|e)$ using **Bayes's Theorem** and then compare the two by dividing the former by the latter:

$$\frac{Pr(h|e \wedge r_n)}{Pr(h|e)} = \frac{Pr(e \wedge r_n|h)}{Pr(e|h)}$$
$$\times \frac{Pr(h)Pr(e|h) + Pr(h')Pr(e|h') + Pr(c)Pr(e|c)}{Pr(h)Pr(e \wedge r_n|h) + Pr(h')Pr(e \wedge r_n|h') + Pr(c)Pr(e \wedge r_n|c)}.$$

Given that our nth means of detecting r is explanatorily discriminating (as stipulated by **RA-Diversity**), we know that $\mathcal{E}(r_n, h|e) \approx 1$ and $\mathcal{E}(\neg r_n, h'|e) \approx 1$. When coupled with **Success**, these considerations entail $Pr(r_n|h \wedge e) \approx 1$ and $Pr(r_n|h' \wedge e) \approx 0$ respectively (Schupbach, 2018, Appendix).[76] From $Pr(r_n|h' \wedge e) \approx 0$, it follows that $Pr(e \wedge r_n|h') = Pr(e|h')Pr(r_n|h' \wedge e) \approx 0$. These likelihoods accordingly reflect Horwich's intuition that diverse (explanatorily discriminating) evidence that favors h over h' should effectively eliminate h':

$$\frac{Pr(h|e \wedge r_n)}{Pr(h|e)} = \frac{Pr(e \wedge r_n|h)}{Pr(e|h)}$$
$$\times \frac{Pr(h)Pr(e|h) + Pr(h')Pr(e|h') + Pr(c)Pr(e|c)}{Pr(h)Pr(e \wedge r_n|h) + \cancel{Pr(h')Pr(e \wedge r_n|h')} + Pr(c)Pr(e \wedge r_n|c)}$$
$$\approx \frac{Pr(e \wedge r_n|h)}{Pr(e|h)} \times \frac{Pr(h)Pr(e|h) + Pr(h')Pr(e|h') + Pr(c)Pr(e|c)}{Pr(h)Pr(e \wedge r_n|h) + Pr(c)Pr(e \wedge r_n|c)}.$$

$Pr(r_n|h \wedge e) \approx 1$ formalizes the idea that h accounts so well for our robustly detecting r using this nth means. Given this likelihood, we can show $Pr(e \wedge r_n|h) = Pr(e|h)Pr(r_n|h \wedge e) \approx Pr(e|h)$. This allows us to simplify the above equation further:

$$\frac{Pr(h|e \wedge r_n)}{Pr(h|e)} \approx \frac{Pr(h)Pr(e|h) + Pr(h')Pr(e|h') + Pr(c)Pr(e|c)}{Pr(h)Pr(e|h) + Pr(c)Pr(e \wedge r_n|c)}. \tag{3.1}$$

Comparing like terms between the denominator and numerator of the right hand side of (3.1) (and remembering that the axioms of probability entail $Pr(e \wedge r_n|c) \leq Pr(e|c)$), we see that this ratio must be top-heavy. Thus, in

---

76 Note that these likelihoods intuitively fit standard cases of RA. E.g., it is highly likely (indeed, nearly certain) that we should witness the Brownian motion using fragments of inorganic matter (under background conditions specifying the proper setup of our experiment) given that the motion is due to unobserved agitations of the fluid medium, but this result is highly unlikely (indeed, it is nearly certain that it will not be observed) given that the motion is due to a vital force in organic matter.

these circumstances, $Pr(h|e \wedge r_n) > Pr(h|e)$; a successful increment of RA will indeed confirm (increase the probability of) target hypothesis h to some extent.

What determines the extent? Examining the right side of (3.1) again, observe that the numerator of this ratio is greater than the denominator *at least* by a difference of $Pr(h')Pr(e|h')$. On the one hand, h''s prior probability $Pr(h')$ measures how plausible h's competitor was to begin with. On the other hand, $Pr(e|h')$ roughly measures competitor h''s fit with the evidence, prior to its being ruled out with the addition of r_n. These factors, considered together, provide an estimate of h''s overall epistemic virtue pre-elimination. So how good was h' (both on its own and in relation to e) before r_n? The answer to this question tells us how much better off (at least) h is now that we've ruled h' out. The target hypothesis h soaks up the epistemic virtue that h' had going for it prior to being eliminated; spoils to the victor.

Consistent Epistemic Competitors

When h and its competitors are mutually exclusive, a successful increment of RA provides confirmation for h by ruling out certain ways in which h could be false. But what about cases in which h and its competitors are consistent with one another, but nonetheless compete to some degree in light of e (à la the above **Competition** condition)? In such a case, by chopping away at h's competitors, explanatorily discriminating evidence also precludes possible ways in which h could be *true*. So it's not obvious that RA will provide an effective strategy for confirming h in these cases.

Allowing for the possibility that h and h' are true together, we expand $Pr(h|e)$ using **Total Probability**:

$$Pr(h|e) = Pr(h|e \wedge h')Pr(h'|e) + Pr(h|e \wedge \neg h')Pr(\neg h'|e). \qquad (3.2)$$

(3.2) represents $Pr(h|e)$ as a weighted average of $Pr(h|e \wedge h')$ and $Pr(h|e \wedge \neg h')$. How do these compare to one another? **Competition** implies that h and h' disconfirm one another conditional on e (§3.3.2), and thus $Pr(h|e \wedge h') < Pr(h|e \wedge \neg h')$.[77] $Pr(h|e \wedge \neg h')$ thus sets a maximum cap (greater than $Pr(h|e \wedge h')$ to the extent that these hypotheses compete) on the value of $Pr(h|e)$.

$Pr(h|e \wedge r_n)$ can be expanded using **Total Probability** as follows:

---

[77] Since h and h' disconfirm one another conditional on e, $Pr(h|e) > Pr(h|e \wedge h')$. But then, since $Pr(h|e)$ is a weighted average of $Pr(h|e \wedge h')$ and $Pr(h|e \wedge \neg h')$, $Pr(h|e \wedge h') < Pr(h|e \wedge \neg h')$.

$$Pr(h|e \wedge r_n) = Pr(h|e \wedge r_n \wedge h')Pr(h'|e \wedge r_n)$$
$$+ Pr(h|e \wedge r_n \wedge \neg h')Pr(\neg h'|e \wedge r_n).$$

Recall from the previous demonstration that **RA-Diversity** and **Success** jointly entail that $Pr(r_n|h' \wedge e) \approx 0$. Consequently, it follows that $Pr(h'|e \wedge r_n) \approx 0$, and so $Pr(\neg h'|e \wedge r_n) \approx 1$.[78] **RA-Diversity** thus again has an eliminative effect in our analysis:

$$Pr(h|e \wedge r_n) = Pr(h|e \wedge r_n \wedge h')Pr(h'|e \wedge r_n)$$
$$+ Pr(h|e \wedge r_n \wedge \neg h')Pr(\neg h'|e \wedge r_n)$$
$$\approx Pr(h|e \wedge r_n \wedge \neg h') \qquad (3.3)$$

Does a successful increment of RA lend confirmation to h in these cases? We may specify the conditions under which it does, in light of the preceding discussion, by comparing (3.2) and (3.3). Whether or not $Pr(h|e \wedge r_n) > Pr(h|e)$ comes down to whether

$$Pr(h|e \wedge r_n \wedge \neg h') > Pr(h|e \wedge h')Pr(h'|e) + Pr(h|e \wedge \neg h')Pr(\neg h'|e). \quad (3.4)$$

Recall that the right side of (3.4) is a weighted average of $Pr(h|e \wedge h')$ and $Pr(h|e \wedge \neg h')$, with a maximum cap value of $Pr(h|e \wedge \neg h')$. The strength of inequality (3.4) can thus be gauged by asking two questions:

1. How much (if at all) greater than the right side's maximum cap $Pr(h|e \wedge \neg h')$ is the left side $Pr(h|e \wedge r_n \wedge \neg h')$?
2. How much (if at all) greater than the weighted average is this average's maximum cap; that is, for fixed weights $Pr(h'|e)$ and $Pr(\neg h'|e)$, to what extent is $Pr(h|e \wedge \neg h')$ greater than $Pr(h|e \wedge h')$?

The degree to which r_n confirms h is thus determined by the strength of the following two inequalities: $Pr(h|e \wedge \neg h') > Pr(h|e \wedge h')$ and $Pr(h|e \wedge r_n \wedge \neg h') > Pr(h|e \wedge \neg h')$. This result can be intuitively interpreted and motivated. The first of these inequalities corresponds directly to our explication of the degree to which h_1 and h_2 compete in light of e (see §3.3.2 and footnote 77). And this particular inequality corresponds with the degree to which the presence of h' disconfirms h conditional on e. The second inequality directly corresponds with the degree to which the new detection r_n confirms h, in light of the past evidence and h''s being eliminated. The degree to which an increment of RA supports h then is determined by joining the degree to which h'

---

[78] Given that $Pr(h'|e \wedge r_n) = Pr(h' \wedge e)Pr(r_n|h' \wedge e)/Pr(e \wedge r_n)$.

was (pre-elimination) competing with/disconfirming h to the additional degree of confirmation that r_n lends to h in light of h''s elimination.

3.4 Simplicity and Evidential Fit

Thus far, this section has discussed some of the factors considered by scientists when evaluating how well their hypotheses relate to evidence. Explanatory hypotheses are often evaluated by how expected they make the salient evidence. And the same observations across diverse means of detection can provide especially strong support for hypotheses that would lead us to expect such robust results.

Scientists, however, don't only consider the relation between hypotheses and evidence when reasoning theoretically. In fact, one of the most venerable methodological principles employed by scientists (and others, including philosophers and theologians) throughout the ages focuses on a monadic property of hypotheses: simplicity.[79] The simplicity of a hypothesis, in one important sense,[80] refers to its reluctance to say too much. This notion of simplicity corresponds with classic statements of "Ockham's razor." For example, Duns Scotus (c. 1265–1308) advises that "we should always posit fewer things when the appearances can be saved thereby [...] in positing more things we should always indicate the manifest necessity on account of which so many things are posited" (1998, p. 349). And William of Ockham (c. 1287–1347) asserts that "it is futile to do with more what can be done with fewer" (1986, pp. 157–158).

To be sure, such statements of Ockham's Razor make implicit (or sometimes explicit) reference to some body of evidence. While this observation might suggest that simplicity is a relational property after all, one might instead interpret such principles as relating the monadic property of simplicity to the evidence. Hypotheses are simple to the extent that they postulate less. But such simplicity is only desirable to the extent that we retain a sufficient degree of evidential fit. That is, hypotheses should say as little as is needed still to accommodate the evidence sufficiently well. This observation points to a trade-off that scientists routinely must navigate.

---

[79] Some of the research presented here was carried out in collaboration with David H. Glass.

[80] Formal work on simplicity has done much to disentangle different concepts of simplicity (Good, 1968; Forster and Sober, 1994; Kelly, 2011; Sober, 2015). The concept that I have in mind corresponds with Good's informational notion of simplicity, "the reciprocal of complexity" (1968, pp. 139–140), which e.g. decreases with the logical strength of one's hypothesis.

Consider an example of this trade-off at work. Scientists consider numerous explanatory hypotheses regarding the mass extinction event at the Cretaceous–Paleogene (K–Pg) boundary (responsible for the extinction of the dinosaurs). These include bolide impact, massive volcanic activity and flooding, climate change, and sea level regression. Some geologists have argued for bolide impact as an informationally simple, "smoking gun" explanation (Cleland, 2011), sufficient on its own (i.e., without committing one way or the other to other postulated historical causes) to account adequately for the variety of historical traces and evidence relevant to the event. Cited evidence includes, in particular, anomalously high levels of iridium in deep-sea limestones dating from the K–Pg boundary (Alvarez et al., 1980), the Chicxulub crater (Hildebrand et al, 1991), and "ejecta-rich deposits" in distribution patterns related to distance from the crater (Schulte et al., 2010).

More recent work has focused on the question of whether we don't explain more evidence at a deeper level by logically strengthening this position, opting for an explanatory hypothesis citing impact in conjunction with other of the above causal factors (Archibald et al., 2010; Courtillot and Fluteau, 2010; Keller et al., 2010; Renne et al., 2015). For example, Archibald et al. suggest that the explanans that commits only to the impact hypothesis fails adequately to fit the evidence found in "countless studies of how vertebrates and other terrestrial and marine organisms fared at the end of the Cretaceous." In order to fit more evidence adequately, it's common to find scientists opting for a more complex explanans – e.g. one that conjoins impact with volcanic flooding.

In such cases, when should one opt for an informationally simpler hypothesis? One prima facie plausible answer might be to go as simple as fit allows; that is, go for the simplest hypothesis that still fits the evidence as well as any more complex hypothesis. This answer gives evidential fit lexical priority over simplicity. However, this idea would often enjoin reasoners to "throw the kitchen sink" at any body of evidence. In the above example, with so many potentially causal factors available, each of which fits distinct facets of the evidence to varying degrees, insofar as we can make even slight improvements in accounting for the evidence by postulating all of the above factors, this answer would encourage us to do so! But that's clearly not wise. At some point, we postulate too much by only fitting the evidence somewhat better while significantly claiming more and thus standing a substantially higher chance of being wrong; at some point, such complex hypotheses start to look ad hoc.

The relation between simplicity and evidential fit rather seems to be a genuine trade-off. Some cost in informational complexity (postulating strictly more) can be worth consequent gains in evidential fit. But it's also the case that some gains in informational simplicity (postulating strictly less) may be worth consequent costs in evidential fit. Here, we explore how the Bayesian machinery can help guide one in appropriately navigating this trade-off.

3.4.1 A Bayesian Approach to Simplicity and Fit

A simpler hypothesis, in our sense, commits to strictly less information about the world. Hence, we may frame simpler versus more complex options as the choice between logically asymmetrical options h_1 and $h_1 \wedge h_2$. The single conjunct option h_1 is uncommitted with respect to h_2; it is the option of only committing with respect to h_1 and so is rightly thought of as equivalent to $h_1 \wedge (h_2 \vee \neg h_2)$. It is *not* the committed stance characterized by the conjunction $h_1 \wedge \neg h_2$, which is on a logical par with $h_1 \wedge h_2$.[81]

As such, a hypothesis's simplicity is reflected in its probability, since logically weaker propositions are more probable than their logically stronger counterparts (recall **Entailment**, from §1.3.3). Differences in simplicity (in our sense and relative to the contexts we're interested in) thus correspond to differences in probability: $Pr(h_1) \geq Pr(h_1 \wedge h_2)$. And indeed, a hypothesis's prior probability itself can be thought of as a measure of that hypothesis's degree of simplicity for the sake of such comparisons.

Complexity, as the inverse of simplicity, can be given a similarly basic explication. In fact, in work that develops Shannon's (1948) communication theory along Bayesian lines, Bar-Hillel and Carnap (1953) and Good (1966, 1968) propose the log-normalized reciprocal of a hypothesis's prior probability as its measure of "complexity" or "amount of semantic information":

$$I(h) = -\log Pr(h) = \log \frac{1}{Pr(h)}.$$

Regardless of I's plausibility as a general measure of semantic information, it works manifestly well for our comparisons of complexity across logically dependent propositions.

The explication of an appropriate notion of evidential fit for our question is somewhat more complicated, but the measure of power $\mathcal{E}_{SS}$ (§3.2.2) provides a fruitful launching point. As noted earlier, this notion of power has to do with a hypothesis's ability to increase the degree to which we expect the explanandum or evidence. As one could argue by way of the K–Pg example above, one compelling reason why one might be willing to opt for a simpler explanation

---

81   The advice to take h_1 over $h_1 \wedge h_2$ thus corresponds to Sober's "razor of silence," as opposed to his "razor of denial" (2009, §5; 2015, p. 12).

is if doing so purchases greater power in this sense; that is, if by doing so, the explanandum becomes more expected.

However, the notion of evidential fit that gets traded off with considerations of simplicity evidently goes beyond mere comparisons of power. A familiar, major incentive for logically strengthening one's hypothesis is to account for a *wider* array of evidence. Comparisons of scope hold considerable sway, for example, in the K–Pg case where different causal factors are individually better aligned with different parts of the evidence. But in an important family of cases, $\mathcal{E}_{SS}$ fails to reflect the virtue of scope. If h entails two distinct (and logically independent) explananda individually: $h \vDash e_1$, $h \vDash e_2$, h has equal (and maximal) power over e_1, e_2, *and* $e_1 \wedge e_2$; that is, $\mathcal{E}_{SS}(e_1, h) = \mathcal{E}_{SS}(e_2, h) = \mathcal{E}_{SS}(e_1 \wedge e_2, h)$. There is no gain in power by accounting for $e_1 \wedge e_2$ as opposed to accounting only for e_1 (or e_2). What one wants, in order to account for the benefit of conjunctive explanations, is a measure that incorporates the virtue of power while also reflecting the virtue of scope under such conditions.

This is exactly what is provided by the following measure of explanatoriness, defended by Good (1960, 1966, 1968)[82] and more recently by McGrew (2003):

$$\mathcal{E}_{GM}(e, h) = \log \left(\frac{Pr(e|h)}{Pr(e)} \right).$$

Like $\mathcal{E}_{SS}$, $\mathcal{E}_{GM}$ reflects considerations of power.[83] But unlike $\mathcal{E}_{SS}$ in the preceding formal situation involving scope, $\mathcal{E}_{GM}$ has the following desirable implication:

$$\mathcal{E}_{GM}(e_1 \wedge e_2, h) = \log \left(\frac{Pr(e_1 \wedge e_2|h)}{Pr(e_1 \wedge e_2)} \right) > \log \left(\frac{Pr(e_1|h)}{Pr(e_1)} \right)$$
$$= \mathcal{E}_{GM}(e_1, h).$$

Thus, $\mathcal{E}_{GM}$ provides an explication of evidential fit that simultaneously captures considerations of power and scope. At the same time, this measure has normative implications mirroring $\mathcal{E}_{SS}$'s (cf., §3.2.2) for situations in which a hypothesis scores positively with respect to the evidence:

$$\mathcal{E}_{GM}(e, h) > 0 \Leftrightarrow Pr(e|h) > Pr(e)$$
$$\Leftrightarrow Pr(e|h) > Pr(e|\neg h) \tag{L}$$
$$\Leftrightarrow Pr(h|e) > Pr(h). \tag{C}$$

---

[82] Good (1968, p. 126) provides an information-theoretic interpretation of this measure as "the amount of information concerning e provided by h."

[83] As Schupbach and Sprenger (2011, §3.1) show, however, $\mathcal{E}_{GM}$ has a property that arguably rules it out as an adequate *measure* of power. Specifically, $\mathcal{E}_{GM}(e, h) = \mathcal{E}_{GM}(e \wedge e', h)$ for any e, h, and e' such that e' is statistically independent of h conditional on e.

And like $\mathcal{E}_{SS}$, comparisons of hypotheses with respect to the same evidence using $\mathcal{E}_{GM}$ amount to likelihood comparisons (cf., §3.2.3):

$$\mathcal{E}_{GM}(e, h_1) = \log\left(\frac{Pr(e|h_1)}{Pr(e)}\right) > \log\left(\frac{Pr(e|h_2)}{Pr(e)}\right) = \mathcal{E}_{GM}(e, h_2)$$
$$\Leftrightarrow Pr(e|h_1) > Pr(e|h_2).$$

Explicating degree of evidential fit using $\mathcal{E}_{GM}$, we can lend formal backing to our reasons against the proposal of "going as simple as fit allows." Giving evidential fit lexical priority over simplicity would amount here to penalizing for complexity $I(h)$ only in cases where $\mathcal{E}_{GM}$ does not already favor a more complex hypothesis. But, because comparisons of $\mathcal{E}_{GM}$ (apropos the same e) reduce to likelihood comparisons, this idea would render simpler hypotheses virtually impossible to justify. Let h_1 be a hypothesis with some degree of evidential fit $\mathcal{E}_{GM}(e, h_1)$. Now consider any additional h_2 at all; so long as it isn't contrary to h_1, it can be irrelevant to or as negatively associated with h_1 as you like. If e is even slightly more likely given $h_1 \wedge h_2$ than it is given h_1 alone, this account tells us to favor the complex, conjunctive hypothesis. So long as we increase the likelihood by doing so, this proposal would encourage us to "throw the kitchen sink" at e.

3.4.2 Striking a Balance

Can Bayesianism guide us in striking a balance between a hypothesis's simplicity and its evidential fit? Interestingly, **Bayes's Theorem** can be represented in a way that suggests it does just this. This can be seen more readily by considering the logarithm of **Bayes's Theorem**:

$$\log Pr(h|e) = \log\left(\frac{Pr(h)Pr(e|h)}{Pr(e)}\right)$$
$$= \log\left(\frac{Pr(e|h)}{Pr(e)}\right) - \log\left(\frac{1}{Pr(h)}\right)$$
$$= \mathcal{E}_{GM}(e, h) - I(h).$$

By simply comparing hypotheses via **Bayes Theorem**, one automatically accounts for that hypothesis's fit with the evidence, while simultaneously penalizing for the complexity of the hypothesis.

A moment's thought, however, is all it takes to realize that **Bayes's Theorem** certainly does *not* strike the right balance, and this for instructive reasons. After all, this theorem leads to the following criterion for accepting more complex hypotheses:

Posteriors. The more complex hypothesis $h_1 \wedge h_2$ is preferable to the simpler hypothesis h_1 iff $\mathcal{E}_{GM}(e, h_1 \wedge h_2) - I(h_1 \wedge h_2) > \mathcal{E}_{GM}(e, h_1) - I(h_1) \Leftrightarrow Pr(h_1 \wedge h_2 | e) > Pr(h_1 | e)$.

We concluded §3.4.1 by noting that simpler hypotheses would be virtually impossible to justify if we gave $\mathcal{E}_{GM}$ lexical priority over simplicity (i.e., penalties for $I(h)$). But here we see that complex explanations would be outrightly impossible to justify according to **Posteriors** (by **Entailment**, §1.3.3). We could weaken this condition to $Pr(h_1 \wedge h_2 | e) \geq Pr(h_1 | e)$, thus at least making it possible to satisfy. However, this weakened criterion is satisfied if and only if $Pr(h_1 \wedge h_2 | e) = Pr(h_1 | e)$ and thus $Pr(h_2 | h_1 \wedge e) = 1$; this amounts to requiring that h_2 adds no informational complexity at all to h_1 in the light of e, and thus $h_1 \wedge h_2$ is only more complex in form and not in informational content than h_1. The upshot is that **Posteriors** describes a condition under which simplicity gets absolute priority over evidential fit. More complex hypotheses can never be favored, and we're ultimately always instructed to opt for tautologies. In response to our central question, "When should one opt for an informationally simpler hypothesis?", this account responds "always!"

If we let $\mathcal{E}_{GM}$ have lexical priority over simplicity, we effectively ignore simplicity altogether, always favoring more complex hypotheses. By contrast, **Posteriors** places extreme weight on considerations of simplicity, banning more complex hypotheses altogether. The proper balance must reside somewhere between these extreme options.

Interestingly, Good (1968) himself highlighted the fact that his measure $\mathcal{E}_{GM}$ ignores considerations of informational complexity. Correcting for this, he proposes the following family of "strong" measures of explanatoriness, $\mathcal{E}_{\gamma}$ (where $0 < \gamma < 1$):

$$\mathcal{E}_{\gamma}(e, h) = \log \left(\frac{Pr(e|h) \cdot Pr(h)^{\gamma}}{Pr(e)} \right)$$
$$= \log \left(\frac{Pr(e|h)}{Pr(e)} \right) + \gamma \log Pr(h)$$
$$= \log \left(\frac{Pr(e|h)}{Pr(e)} \right) - \gamma \log \left(\frac{1}{Pr(h)} \right). \tag{3.5}$$

Note, on the one hand, that if $\gamma = 0$, then the resulting measure reduces to $\mathcal{E}_{\gamma=0} = \mathcal{E}_{GM}$, with the result that complexity is given no weight. On the other hand, if $\gamma = 1$, then the resulting measure is (the logarithm of) **Bayes's Theorem**, with the aforementioned result that complexity is given absolute priority. Since $0 < \gamma < 1$, this family of measures provides any number of more plausible accounts of how to balance fit with simplicity.

How might we choose a particular measure (value of γ) out of this family? One way is to think about the condition under which this measure should zero out. Glass (2021) defends a particular measure in this way, making important connections to Bayesian information theory. Here, we may sidestep those connections and provide the following similar argument. The central thesis is that $\mathcal{E}_\gamma$, being a measure that balances considerations of evidential fit with complexity, should zero out exactly when the payoff a hypothesis brings in terms of its evidential fit, $\mathcal{E}_{GM}$, perfectly balances with the cost it *ultimately* incurs in terms of complexity – that is, its complexity in the light of e, or $I(h|e) = -\log Pr(h|e)$:[84]

$$\mathcal{E}_{GM}(e, h) = \log\left(\frac{Pr(e|h)}{Pr(e)}\right) = \log\left(\frac{1}{Pr(h|e)}\right) = I(h|e)$$

$$\Leftrightarrow \log\left(\frac{Pr(e|h)}{Pr(e)}\right) = \log\left(\frac{Pr(e)}{Pr(h)Pr(e|h)}\right)$$

$$\Leftrightarrow \log Pr(e|h) - \log Pr(e) = \log Pr(e) - \log Pr(h)$$
$$- \log Pr(e|h)$$

$$\Leftrightarrow 2[\log Pr(e|h) - \log Pr(e)] = -\log Pr(h)$$

$$\Leftrightarrow \log\left(\frac{Pr(e|h)}{Pr(e)}\right) = {}^{1}/{}_{2} \cdot \log\left(\frac{1}{Pr(h)}\right).$$

Referring to (3.5) and setting $\mathcal{E}_\gamma(e, h) = 0$, this straightforwardly implies that $\gamma = 1/2$. This result provides formal backing for Good's (1968, p. 130) remark that setting $\gamma = 1/2$ "gives equal weights to [evidential fit and complexity]." The result is the following Bayesian measure, which provides the desired guidance regarding how to trade off evidential fit with simplicity:

$$\mathcal{E}_{\gamma=.5}(e, h) = \log\left(\frac{P(e|h) P(h)^{1/2}}{P(e)}\right).$$

Since Good and Glass (2021) both have previously proposed and defended this account, we refer to it as $\mathcal{E}_{GG}$.

$\mathcal{E}_{GG}$ inspires the following criterion for accepting simpler hypotheses:

Simplicity-Fit. The simpler hypothesis h_1 is preferable to the more complex $h_1 \wedge h_2$ iff $\mathcal{E}_{GG}(e, h_1) > \mathcal{E}_{GG}(e, h_1 \wedge h_2)$.

---

[84] Prima facie, one might instead think that $\mathcal{E}_\gamma(e, h) = 0$ when the payoff a hypothesis brings in terms of its evidential fit perfectly balances with its complexity simpliciter $I(h)$. However, this requirement would again lead to the problematic **Posteriors** account, since it implies that $\gamma = 1$. Furthermore, since these trade-offs are managed relative to a given e, it only makes sense to evaluate h's complexity with the supposition of e in place.

This criterion implies the following more comprehensible answer to our central question of when one ought to opt for an informationally simpler hypothesis:[85]

$$\mathcal{E}_{GG}(e, h_1) > \mathcal{E}_{GG}(e, h_1 \wedge h_2)$$

$$\Leftrightarrow \log\left(\frac{1}{Pr(h_2|e \wedge h_1)}\right) > \log\left(\frac{Pr(e|h_1 \wedge h_2)}{Pr(e|h_1)}\right) \tag{3.6}$$

$$\Leftrightarrow I(h_2|e \wedge h_1) > \mathcal{E}_{GM}(e, h_2|h_1).$$

That is, a simpler hypothesis h_1 should be preferred over the complex $h_1 \wedge h_2$ when the informational complexity that would be added by additionally postulating h_2 outweighs the increase in evidential fit this additional postulate brings about. In other words, the improved fit is not worth the cost in complexity. In the opposite situation then, one ought to opt for a more complex hypothesis $h_1 \wedge h_2$ when $\mathcal{E}_{GG}(e, h_2|h_1) > I(h_2|e \wedge h_1)$, or in words, when postulating h_2 in addition to h_1 provides an increase in evidential fit that is worth the inevitable, corresponding cost in complexity.

3.4.3 Application

In order to evaluate whether the above account does indeed capture the considerations raised by scientists when balancing simplicity against fit, let's return to the K–Pg case. Recall that scientists have recently debated whether we should logically strengthen our explanans, conjoining the already favored impact postulate with other causal factors, in order to account for more of the available evidence. Archibald et al. (2010, p. 973), for example, write:

> A simplistic extinction scenario has not stood up to the countless studies of how vertebrates and other terrestrial and marine organisms fared at the end of the Cretaceous. Patterns of extinction and survival were varied, pointing to

---

[85] Proof. $\mathcal{E}_{GG}(e, h_1) > \mathcal{E}_{GG}(e, h_1 \wedge h_2)$

$$\Leftrightarrow \frac{Pr(e|h_1)Pr(h_1)^{1/2}}{Pr(e)} > \frac{Pr(e|h_1 \wedge h_2)Pr(h_1 \wedge h_2)^{1/2}}{Pr(e)}$$

$$\Leftrightarrow \left(\frac{Pr(h_1)}{Pr(h_1 \wedge h_2)}\right)^{1/2} > \frac{Pr(e|h_1 \wedge h_2)}{Pr(e|h_1)}$$

$$\Leftrightarrow \frac{Pr(h_1)}{Pr(h_1 \wedge h_2)} > \frac{Pr(e|h_1 \wedge h_2)Pr(e|h_1 \wedge h_2)}{Pr(e|h_1)Pr(e|h_1)}$$

$$\Leftrightarrow \frac{Pr(h_1)Pr(e|h_1)}{Pr(h_1 \wedge h_2)Pr(e|h_1 \wedge h_2)} > \frac{Pr(e|h_1 \wedge h_2)}{Pr(e|h_1)}$$

$$\Leftrightarrow \frac{Pr(e \wedge h_1)}{Pr(e \wedge h_1 \wedge h_2)} > \frac{Pr(e|h_1 \wedge h_2)}{Pr(e|h_1)}$$

$$\Leftrightarrow \log\left(\frac{1}{Pr(h_2|e \wedge h_1)}\right) > \log\left(\frac{Pr(e|h_1 \wedge h_2)}{Pr(e|h_1)}\right) \quad \square$$

multiple causes at this time – including impact, marine regression, volcanic activity, and changes in global and regional climatic patterns.

According to this line of argument, we need to postulate multiple causal factors to account for the complex evidence of this case. And the difference is captured with a simple appeal to the power- and scope-incorporating notion of evidential fit. Letting E represent the complex conjunction of evidence and focusing for simplicity on the bolide impact H_1 and volcanic flooding H_2 hypotheses, we may explicate the conclusion for which these scientists are arguing as:

$$\frac{Pr(E|H_1 \wedge H_2)}{Pr(E)} \gg \frac{Pr(E|H_1)}{Pr(E)}.$$

Of course, our account requires that more be said at this point; after all, if this were the sole desideratum for deciding whether to strengthen one's hypothesis, there would seem to be no end to the additional H_i's we could introduce that would still further boost E's likelihood (conditional on the hypotheses so far accepted). Specifically, our account proceeds to check whether such increased fit with the evidence is worth the price in complexity paid by logically strengthening the position.

It's thus reassuring to find scientists debating exactly this further point. For example, Renne et al. (2015) write:

> Bolide impact and flood volcanism compete as leading candidates for the cause of terminal-Cretaceous mass extinctions. High-precision argon-argon data indicate that these two mechanisms may be genetically related, and neither can be considered in isolation. The existing Deccan Traps magmatic system underwent a state shift approximately coincident with the Chicxulub impact and the K-Pg mass extinctions [...] Initiation of this new regime occurred within ∼50,000 years of the impact, which is consistent with transient effects of impact-induced seismic energy.

Renne et al.'s point is distinct from the previous one. Rather than pointing to evidence calling for a more complex explanans, they argue that H_1 and H_2 are themselves causally related. That is, they provide reason to think that a bolide impact would kick off massive volcanic activity. To the extent that they establish this point, volcanic flooding should be expected to occur on the heels of a bolide impact (conditional or not on E), and so $Pr(H_2|E \wedge H_1) \gg Pr(H_2|E)$. This inequality directly bears on the informational complexity added to an explanation by additionally postulating H_2, assuming that the explanans already contains H_1 (conditional or not on E), for it shows that the complexity $I(H_2|E \wedge H_1)$ cannot be substantial.

Recalling result (3.6) above, our account clarifies that the more complex explanatory hypothesis ought to be preferred in this case if and only if:

$$\log \left(\frac{Pr(E|H_1 \wedge H_2)}{Pr(E|H_1)} \right) > \log \left(\frac{1}{Pr(H_2|E \wedge H_1)} \right).$$

The purpose in discussing this example is not to argue that this inequality is satisfied, a question best left to the scientists. But the point *is* that this is an accurate explication of what the scientists are trying to determine. Our account captures the relevance for this question of exactly those considerations that the scientists are raising. Those arguing that the complex explanans is better in this case are putting forward two distinct arguments. The first amounts to the argument that the left-hand term of the above inequality is substantial; the second amounts to arguing that the right-hand term is not. To the extent that these two arguments are compelling, our account agrees that this would be a case where one should prefer the more complex explanatory hypothesis.

References

Alvarez, L. W., Alvarez, W., Asaro, F., and Michel, H. V. (1980). Extraterrestrial cause for the Cretaceous-Tertiary extinction. *Science*, 208(4448):1095–1108.

Archibald, J. D. et al., (2010). Cretaceous extinctions: multiple causes. *Science*, 328(5981):973.

Bacchus, F. (1990). *Representing and Reasoning with Probabilistic Knowledge: A Logical Approach to Probabilities*. Massachusetts Institute of Technology Press, Cambridge, MA.

Bar-Hillel, Y. and Carnap, R. (1953). Semantic information. *The British Journal for the Philosophy of Science*, 4(14):147–157.

Barnes, J., ed. (1984). *The Complete Works of Aristotle: The Revised Oxford Translation*. Princeton University Press, Princeton, NJ, 6th edition.

Bayes, T. (1763). An essay towards solving a problem in the doctrine of chances. *Philosophical Transactions of the Royal Society*, 53:370–418.

Belot, G. (2013). Bayesian orgulity. *Philosophy of Science*, 80(4):483–503.

Bertrand, J. (1889). *Calcul des Probabilités*. Gauthier-Villars, Paris.

Birnbaum, A. (1962). On the foundations of statistical inference. *Journal of the American Statistical Association*, 57(298):269–306.

Blackwell, D. and Dubins, L. (1962). Merging of opinions with increasing information. *The Annals of Mathematical Statistics*, 33(3):882–886.

Bolzano, B. (1837). *Theory of Science*. Blackwell, Oxford, 1972 edition. Translated by Rolf George.

Boole, G. (1854). *An Investigation of the Laws of Thought: on Which Are Founded the Mathematical Theories of Logic and Probabilities*. Macmillan and Co., London.

Bovens, L. and Hartmann, S. (2003). *Bayesian Epistemology*. Oxford University Press, New York.

Bradley, S. and Steele, K. (2014). Uncertainty, learning, and the "problem" of dilation. *Erkenntnis*, 79(6):1287–1303.

Brössel, P. (2015). On the role of explanatory and systematic power in scientific reasoning. *Synthese*, 192(12):3877–3913.

Brown, R. (1828). A brief account of microscopical observations on the particles contained in the pollen of plants; and on the general existence of active molecules in organic and inorganic bodies. *Edinburgh New Philosophical Journal*, 5:358–371. Page references are to Brown's unpublished report.

Buchak, L. (2013). Belief, credence, and norms. *Philosophical Studies*, 169(2):285–311.

Calcott, B. (2011). Wimsatt and the robustness family: review of Wimsatt's *Re-engineering Philosophy for Limited Beings*. *Biology & Philosophy*, 26(2):281–293.

Carnap, R. (1947). On the application of inductive logic. *Philosophy and Phenomenological Research*, 8(1):133–148.

Carnap, R. (1955). *Statistical and Inductive Probability*. The Galois Institute of Mathematics and *Art*, Brooklyn, New York, pages 279–287.

Carnap, R. (1962). *Logical Foundations of Probability*. University of Chicago Press, Chicago, 2nd edition.

Cartwright, N. (1991). Replicability, reproducibility, and robustness: comments on Harry Collins. *History of Political Economy*, 23(1):143–155.

Chesterton, G. K. (1908). *Orthodoxy*. John Lane Company, New York.

Christensen, D. (1996). Dutch-book arguments depragmatized: epistemic consistency for partial believers. *The Journal of Philosophy*, 93(9):450–479.

Cleland, C. E. (2011). Prediction and explanation in historical natural science. *British Journal for the Philosophy of Science*, 62(3):551–582.

Cohen, M. P. (2015). On Schupbach and Sprenger's measures of explanatory power. *Philosophy of Science*, 82(1):97–109.

Cohen, M. P. (2016). On three measures of explanatory power with axiomatic representations. *The British Journal for the Philosophy of Science*, 67(4):1077–1089.

Colyvan, M. (2004). The philosophical significance of Cox's theorem. *International Journal of Approximate Reasoning*, 37(1):71–85.

Colyvan, M. (2008). Is probability the only coherent approach to uncertainty? *Risk Analysis*, 28(3):645–652.

Colyvan, M. (2013). Idealisations in normative models. *Synthese*, 190(8): 1337–1350.

Courtillot, V. and Fluteau, F. (2010). Cretaceous extinctions: the volcanic hypothesis. *Science*, 328(5981):973–974.

Cox, R. T. (1946). Probability, frequency and reasonable expectation. *American Journal of Physics*, 14(1):1–13.

Cox, R. T. (1961). *The Algebra of Probable Inference*. The John Hopkins University Press, Baltimore, MD.

Crupi, V., Fitelson, B., and Tentori, K. (2008). Probability, confirmation, and the conjunction fallacy. *Thinking and Reasoning*, 14(2):182–199.

Crupi, V. and Tentori, K. (2012). A second look at the logic of explanatory power (with two novel representation theorems). *Philosophy of Science*, 79(3):365–385.

Crupi, V., Tentori, K., and Gonzalez, M. (2007). On Bayesian measures of evidential support: theoretical and empirical issues. *Philosophy of Science*, 74(2):229–252.

Culp, S. (1994). Defending robustness: the bacterial mesosome as a test case. *PSA: Proceedings of the Biennial Meeting of the Philosophy of Science Association*, 1: Contributed Papers:46–57.

Darwin, C. (1872). *The Origin of Species*. John Murray, London, 6th edition. Page references are to the Modern Library edition, 1998.

de Finetti, B. (2017). *Theory of Probability: A Critical Introductory Treatment*. Wiley Series in Probability and Statistics. John Wiley & Sons, New York.

Douven, I. (2013). Inference to the Best Explanation, Dutch Books, and inaccuracy minimisation. *The Philosophical Quarterly*, 63(252):428–444.

Douven, I. (2020). The ecological rationality of explanatory reasoning. *Studies in History and Philosophy of Science Part A*, 79(Feb.):1–14.

Duns Scotus, J. (1998). *Questions of the Metaphysics of Aristotle*, Text Series, no. 19, vol. 2. Franciscan Institute Publishers, St. Bonaventure, NY.

Earman, J. (1992). *Bayes or Bust? A Critical Examination of Bayesian Confirmation Theory*. Massachusetts Institute of Technology Press, Cambridge, MA.

Easwaran, K. (2016). Conditional probability. In Hájek, A. and Hitchcock, C., eds., *The Oxford Handbook of Probability and Philosophy*, chapter 9, pages 167–182. Oxford University Press, Oxford.

Easwaran, K. (2019). Conditional probabilities. In Pettigrew, R. and Weisberg, J., eds., *The Open Handbook of Formal Epistemology*, pages 131–198. PhilPapers Foundation. ISBN 978-1-9991763-0-3

Edwards, A. W. F. (1992). *Likelihood: An Account of the Statistical Concept of Likelihood and Its Application to Scientific Inference*. The John Hopkins University Press, Baltimore, expanded edition. First edition published 1972.

Einstein, A. and Infeld, L. (1938). *The Evolution of Physics: From Early Concepts to Relativity and Quanta*. Simon & Schuster, New York.

Feldman, M. (1991). On the generic nonconvergence of Bayesian actions and beliefs. *Economic Theory*, 1(4):301–321.

Fitelson, B. (1996). Wayne, Horwich, and evidential diversity. *Philosophy of Science*, 63(4):652–660.

Fitelson, B. (1999). The plurality of Bayesian measures of confirmation and the problem of measure sensitivity. *Philosophy of Science*, 66(Supp.):S362–S378.

Fitelson, B. (2001a). A Bayesian account of independent evidence with applications. *Philosophy of Science*, 68(3):S123–S140.

Fitelson, B. (2001b). Studies in Bayesian Confirmation Theory. PhD thesis, University of Wisconsin – Madison.

Forster, M. R. and Sober, E. (1994). How to tell when simpler, more unified, or less ad hoc theories will provide more accurate predictions. *British Journal for the Philosophy of Science*, 45:1–35.

Gaifman, H. and Snir, M. (1982). Probabilities over rich languages, testing and randomness. *The Journal of Symbolic Logic*, 47(3):495–548.

Garber, D. (1983). Old evidence and logical omniscience in Bayesian confirmation theory. In Earman, J., ed., *Minnesota Studies in the Philosophy of Science*, vol. X, pages 99–131. University of Minnesota Press, Minneapolis.

Garwood, C. (2008). *Flat Earth: The History of an Infamous Idea*. Thomas Dunne Books, New York.

Glass, D. H. (2012). Inference to the Best Explanation: does it track truth? *Synthese*, 185:411–427.

Glass, D. H. (2016). Science, God and Ockham's razor. *Philosophical Studies*, 174(5):1145–1161

Glass, D. H. (2021). Information and explanatory goodness. Unpublished manuscript.

Glymour, C. (1980). *Theory and Evidence*. Princeton University Press, Princeton.

Good, I. J. (1960). Weight of evidence, corroboration, explanatory power, information and the utility of experiments. *Journal of the Royal Statistical Society. Series B (Methodological)*, 22(2):319–331.

Good, I. J. (1966). A derivation of the probabilistic explication of information. *Journal of the Royal Statistical Society: Series B (Methodological)*, 28(3):578–581.

Good, I. J. (1968). Corroboration, explanation, evolving probability, simplicity and a sharpened razor. *British Journal for the Philosophy of Science*, 19(2):123–143.

Hacking, I. (1965). *Logic of Statistical Inference*. Cambridge University Press, Cambridge.

Hacking, I. (1983). *Representing and Intervening*. Cambridge University Press, Cambridge.

Hacking, I. (2006). *The Emergence of Probability: A Philosophical Study of the Early Ideas About Probability, Induction, and Statistical Inference*. Cambridge University Press, Cambridge, 2nd edition.

Haenni, R., Romeijn, J.-W., Wheeler, G., and Williamson, J. (2010). *Probabilistic Logic and Probabilistic Networks*. The Synthese Library. Springer, Dordrecht.

Hájek, A. (2003). What conditional probability could not be. *Synthese*, 137:273–323.

Hájek, A. (2009). What conditional probability also could not be: reply to Kenny Easwaran. Unpublished.

Halpern, J. Y. (1999a). A counterexample to theorems of Cox and Fine. *Journal of Artificial Intelligence Research*, 10(1):67–85.

Halpern, J. Y. (1999b). Cox's theorem revisited. *Journal of Artificial Intelligence Research*, 11:429–435.

Harman, G. H. (1965). The inference to the best explanation. *Philosophical Review*, 74(1):88–95.

Hempel, C. G. (1945). Studies in the logic of confirmation II. *Mind*, 54(214):97–121.

Hempel, C. G. (1965). Aspects of Scientific Explanation. In *Aspects of Scientific Explanation and Other Essays in the Philosophy of Science*, pages 331–496. Free Press, New York.

Hempel, C. G. and Oppenheim, P. (1948). Studies in the logic of explanation. *Philosophy of Science*, 15(2):135–175.

Hildebrand et al, A. R. (1991). Chicxulub crater: a possible Cretaceous/Tertiary boundary impact crater on the Yucatan Peninsula, Mexico. *Geology*, 19:867–871.

Horgan, T. (2017). Troubles for Bayesian formal epistemology. *Res Philosophica*, 94(2):233–255.

Horwich, P. (1982). *Probability and Evidence*. Cambridge University Press, Cambridge.

Howson, C. (2000). *Hume's Problem: Induction and the Justification of Belief*. Clarendon Press, Oxford.

Howson, C. (2011). Bayesianism as a pure logic of inference. In Bandyopadhyay, P. S. and Forster, M. R., (eds.), *Philosophy of Statistics*, volume 7 of *Handbook of Philosophy of Science*, pages 441–471. Elsevier, Amsterdam.

Howson, C. and Urbach, P. (2006). *Scientific Reasoning: The Bayesian Approach*. Open Court, Peru, IL, 3rd edition.

Jaynes, E. T. (1957). Information theory and statistical mechanics. *Physical Review*, 106(4):620–630.

Jaynes, E. T. (2003). *Probability Theory: The Logic of Science*. Cambridge University Press, Cambridge.

Jeffrey, R. (2004). *Subjective Probability: The Real Thing*. Cambridge University Press, Cambridge.

Jeffrey, R. C. (1983a). Bayesianism with a human face. In Earman, J., (ed.), *Testing Scientific Theories*, volume X, pages 133–156. University of Minnesota Press, Minneapolis.

Jeffrey, R. C. (1983b). *The Logic of Decision*. University of Chicago Press, Chicago, 2nd edition.

Jeffreys, H. (1939). *Theory of Probability*. Clarendon Press, Oxford.

Joyce, J. M. (1998). A nonpragmatic vindication of Probabilism. *Philosophy of Science*, 65(4):575–603.

Joyce, J. M. (2009). Accuracy and Coherence: Prospects for an Alethic Epistemology of Partial Belief. In Huber, F. and Schmidt-Petri, C., (eds.), *Degrees of Belief*, vol. 342 of *Synthese Library*, pages 263–297. Springer, New York.

Joyce, J. M. (2010). A defense of imprecise credences in inference and decision making. *Philosophical Perspectives*, 24(1):281–323.

Keller, G., Adatte, T., Pardo, A. et al. (2010). Cretaceous extinctions: evidence overlooked. *Science*, 328(5981):974–975.

Kelly, K. T. (2011). Simplicity, truth, and probability. In Bandyopadhyay, P. S. and Forster, M. R., (eds.), *Philosophy of Statistics*, vol. 7 of *Handbook of the Philosophy of Science*, pages 983–1024. North-Holland, Amsterdam.

Kelly, K. T., Schulte, O., and Juhl, C. (1997). Learning theory and the philosophy of science. *Philosophy of Science*, 64(2):245–267.

Keynes, J. M. (1921). *A Treatise on Probability*. Macmillan, London.

Kimhi, I. (2018). *Thinking and Being*. Harvard University Press, Cambridge, MA.

Kolmogorov, A. N. (1933). *Foundations of the Theory of Probability*. Chelsea Publishing Company, New York.

Kuorikoski, J., Lehtinen, A., and Marchionni, C. (2010). Economic modelling as robustness analysis. *British Journal for the Philosophy of Science*, 61(3):541–567.

Kyburg, H. E. (1961). *Probability and the Logic of Rational Belief*. Wesleyan University Press, Middletown, CT.

Kyburg, H. E. (1976). Chance. *Journal of Philosophical Logic*, 5(3):355–393.

Kyburg, H. E. (1977). Randomness and the right reference class. *The Journal of Philosophy*, 74(9):501–521.

Kyburg, H. E. (1980). Conditionalization. *The Journal of Philosophy*, 77(2):98–114.

Kyburg, H. E. (1983). Levi, Petersen, and direct inference. *Philosophy of Science*, 50(4):630–634.

Kyburg, H. E. (2003). Don't take unnecessary chances! In Kyburg, H. E. and Thalos, M., (eds.), *Probability Is the Very Guide of Life: The Philosophical Uses of Chance*, pages 277–294. Open Court, La Salle, IL.

Kyburg, H. E. and Teng, C. M. (2001). *Uncertain Inference*. Cambridge University Press, Cambridge.

Laplace, P.-S. (1814). *A Philosophical Essay on Probabilities*. Chapman & Hall, Limited, London.

Leitgeb, H. (2013). Reducing belief simpliciter to degrees of belief. *Annals of Pure and Applied Logic*, 164(12):1338–1389.

Leitgeb, H. and Pettigrew, R. (2010a). An objective justification of Bayesianism I: measuring inaccuracy. *Philosophy of Science*, 77(2):201–235.

Leitgeb, H. and Pettigrew, R. (2010b). An objective justification of Bayesianism II: the consequences of minimizing inaccuracy. *Philosophy of Science*, 77(2):236–272.

Levi, I. (1977). Direct inference. *The Journal of Philosophy*, 74(1):5–29.

Levi, I. (1978). Confirmational conditionalization. *The Journal of Philosophy*, 75(12):730–737.

Levi, I. (1980). *The Enterprise of Knowledge: An Essay on Knowledge, Credal Probability, and Chance*. Massachusetts Institute of Technology Press, Cambridge, MA.

Levi, I. (1981). Direct inference and confirmational conditionalization. *Philosophy of Science*, 48(4):532–552.

Levi, I. (1987). The demons of decision. *The Monist*, 70(2):193–211.

Levi, I. (2004). The logic of consistency and the logic of truth. *Dialectica*, 58(4, Special issue: Ramsey):461–482.

Levins, R. (1966). The strategy of model building in population biology. *American Scientist*, 54(4):421–431.

Lewis, D. (1976). Probabilities of conditionals and conditional probabilities. *The Philosophical Review*, 85(3):297–315.

Lewis, D. (1980). A subjectivist's guide to objective chance. In Jeffrey, R. C., (ed.), *Studies in Inductive Logic and Probability*, pages 263–293. University of California Press, Berkeley.

Lindberg, D. C. (2007). *The Beginnings of Western Science: The European Scientific Tradition in Philosophical, Religious, and Institutional Context, Prehistory to A.D. 1450*. University of Chicago Press, Chicago, 2nd edition.

Lipton, P. (2004). *Inference to the Best Explanation*. Routledge, New York, 2nd edition.

Lloyd, E. A. (2009). Varieties of support and confirmation of climate models. *Proceedings of the Aristotelian Society, Supplementary Volumes*, 83:213–232.

Lloyd, E. A. (2010). Confirmation and robustness of climate models. *Philosophy of Science*, 77(5):971–984.

Maher, P. (1993). *Betting on Theories*. Cambridge University Press, Cambridge.

Mayo, D. G. (1996). *Error and the Growth of Experimental Knowledge.* University of Chicago Press, Chicago.

McGrew, L. (2010). Probability kinematics and probability dynamics. *Journal of Philosophical Research,* 35:89–105.

McGrew, L. (2014). Jeffrey conditioning, rigidity, and the defeasible red jelly bean. *Philosophical Studies,* 168:569–582.

McGrew, L. (2016). Evidential diversity and the negation of H: a probabilistic account of the value of varied evidence. *Ergo,* 3(10):263–292.

McGrew, T. (2003). Confirmation, heuristics, and explanatory reasoning. *British Journal for the Philosophy of Science,* 54(4):553–567.

McMullin, E. (1992). *The Inference That Makes Science.* Marquette University Press, Milwaukee, WI.

of Ockham, W. (1986). *Tractatus de Corpore Christi,* volume X of *Opera Theologica.* Franciscan Institute Publishers, St. Bonaventure, NY.

Parker, W. S. (2011). When climate models agree: the significance of robust model predictions. *Philosophy of Science,* 78(4):579–600.

Pearl, J. (1988). *Probabilistic Reasoning in Intelligent Systems: Networks of Plausible Inference.* Morgan Kaufman, San Francisco.

Pedersen, A. P. and Wheeler, G. (2014). Demystifying dilation. *Erkenntnis,* 79(6):1305–1342.

Peirce, C. S. (1878). Deduction, induction, and hypothesis. *Popular Science Monthly,* 13(Aug.):470–482.

Peirce, C. S. (1883). A theory of probable inference. In *Studies in Logic by Members of the Johns Hopkins University,* pages 126–181. Little, Brown, and Company, Boston, MA.

Peirce, C. S. (1893). Reply to the necessitarians: rejoinder to Dr Carus. *The Monist,* 3(4):526–570.

Peirce, C. S. (1931–5). *The Collected Papers of Charles Sanders Peirce,* vol. I–VI. Harvard University Press, Cambridge, MA.

Peirce, C. S. (1958). *The Collected Papers of Charles Sanders Peirce,* vol. VII–VIII. Harvard University Press, Cambridge, MA.

Perrin, J. (1913). *Les Atomes.* Ox Bow Press, Woodbridge, CT. Translated by D. Ll. Hammick.

Pettigrew, R. (2016). *Accuracy and the Laws of Credence.* Oxford University Press, Oxford.

Pettigrew, R. (2019). Epistemic utility arguments for Probabilism. In Zalta, E. N., editor, *The Stanford Encyclopedia of Philosophy.* Metaphysics Research Lab, Stanford University, winter 2019 edition.

Pettigrew, R. and Titelbaum, M. G. (2014). Deference done right. *Philosophers' Imprint,* 14(35):1–19.

Plutynski, A. (2006). Strategies of model building in population genetics. *Philosophy of Science*, 73(5):755–764.

Pollock, J. L. (1987). Defeasible reasoning. *Cognitive Science*, 11(4):481–518.

Pollock, J. L. (1990). *Nomic Probability and the Foundations of Induction*. Oxford University Press, New York.

Popper, K. R. (1959). *The Logic of Scientific Discovery*. Hutchinson, London.

Priest, G. (1979). The logic of paradox. *Journal of Philosophical Logic*, 8(1):219–241.

Priest, G. (2002). Inconsistency and the empirical sciences. In Meheus, J. (ed.), *Inconsistency in Science*, pages 119–128. Kluwer Academic, Dordrecht.

Priest, G. (2006). *Doubt Truth to Be a Liar*. Oxford University Press, Oxford.

Ramsey, F. P. (1926). Truth and Probability. In Mellor, D. H. (ed.), *Philosophical Papers*, pages 52–94. Cambridge University Press, Cambridge. Edited collection published in 1990.

Reichenbach, H. (1949). *The Theory of Probability*. University of California Press, Berkeley.

Renne, P. R., Sprain, C. J., Richards, M. A. et al. (2015). State shift in Deccan volcanism at the Cretaceous-Paleogene boundary, possibly induced by impact. *Science*, 350(6256):76–78.

Rosenkrantz, R. D. (1981). *Foundations and Applications of Inductive Probability*. Ridgeview Press, Atascadero, CA.

Royall, R. (1997). *Statistical Evidence: A Likelihood Paradigm*. Chapman & Hall, Boca Raton, FL.

Russell, J. B. (1991). *Inventing the Flat Earth: Columbus and Modern Historians*. Praeger, Westport, CT.

Savage, L. J. (1954). *The Foundations of Statistics*. Wiley, New York. Page references are to the 1972 Dover edition.

Schulte et al., P. (2010). The Chicxulub asteroid impact and mass extinction at the Cretaceous-Paleogene boundary. *Science*, 327(5970):1214–1218.

Schupbach, J. N. (2011a). Comparing probabilistic measures of explanatory power. *Philosophy of Science*, 78(5):813–829.

Schupbach, J. N. (2011b). *Studies in the Logic of Explanatory Power*. PhD thesis, University of Pittsburgh, Pittsburgh.

Schupbach, J. N. (2015). Robustness, diversity of evidence, and probabilistic independence. In Mäki, U., Votsis, I., Ruphy, S., and Schurz, G. (ed.) *Recent Developments in the Philosophy of Science: EPSA13 Helsinki*, pages 305–316. Springer, Dordrecht.

Schupbach, J. N. (2016). Competing explanations and explaining-away arguments. *Theology and Science*, 14(3):256–267.

Schupbach, J. N. (2017). Inference to the Best Explanation, cleaned up and made respectable. In McCain, K. and Poston, T. (eds.), *Best Explanations: New Essays on Inference to the Best Explanation*, pages 39–61. Oxford University Press, Oxford.

Schupbach, J. N. (2018). Robustness analysis as explanatory reasoning. *British Journal for the Philosophy of Science*, 69(1):275–300.

Schupbach, J. N. and Glass, D. H. (2017). Hypothesis competition beyond mutual exclusivity. *Philosophy of Science*, 84(5):810–824.

Schupbach, J. N. and Sprenger, J. (2011). The logic of explanatory power. *Philosophy of Science*, 78(1):105–127.

Schupbach, J. N. and Sprenger, J. (2014). Explanatory power and explanatory justice. Unpublished. Available online at http://jonahschupbach.com/documents/EPandEJ.pdf.

Seidenfeld, T. (1979). Why I am not an objective Bayesian: some reflections prompted by Rosenkrantz. *Theory and Decision*, 11:413–440.

Seidenfeld, T. (1986). Entropy and uncertainty. *Philosophy of Science*, 53(4):467–491.

Seidenfeld, T. and Wasserman, L. (1993). Dilation for sets of probabilities. *The Annals of Statistics*, 21(3):1139–1154.

Shannon, C. E. (1948). A mathematical theory of communication. *The Bell System Technical Journal*, 27(July):379–423, 623–656.

Shimony, A. (1985). The status of the principle of maximum entropy. *Synthese*, 63(1):35–53.

Skyrms, B. (1985). Maximum entropy inference as a special case of conditionalization. *Synthese*, 63(1):55–74.

Sober, E. (2008). *Evidence and Evolution: The Logic Behind the Science*. Cambridge University Press, Cambridge.

Sober, E. (2009). Parsimony arguments in science and philosophy – a test case for naturalism. *Proceedings and Addresses of the American Philosophical Association*, 83(2):117–155.

Sober, E. (2015). *Ockham's Razors: A User's Manual*. Cambridge University Press, Cambridge.

Sober, E. (2019). *The Design Argument*. Elements in the Philosophy of Religion. Cambridge University Press, Cambridge.

Sprenger, J. and Hartmann, S. (2019). *Bayesian Philosophy of Science*. Oxford University Press, Oxford.

Stegenga, J. (2009). Robustness, discordance, and relevance. *Philosophy of Science*, 76(5):650–661.

Stegenga, J. and Menon, T. (2017). Robustness and independent evidence. *Philosophy of Science*, 84(3):414–435.

Stolarz-Fantino, S., Fantino, E., Zizzo, D. J., and Wen, J. (2003). The conjunction effect: new evidence for robustness. *The American Journal of Psychology*, 116(1):15–34.

Tentori, K., Crupi, V., Bonini, N., and Osherson, D. (2007). Comparison of confirmation measures. *Cognition*, 103(1):107–119.

Thorn, P. D. (2012). Two problems of direct inference. *Erkenntnis*, 76(3):299–318.

Titelbaum, M. G. (2013). *Quitting Certainties: A Bayesian Framework Modeling Degrees of Belief.* Oxford University Press, Oxford.

van Fraassen, B. C. (1984). Belief and the will. *The Journal of Philosophy*, 81(5):235–256.

van Fraassen, B. C. (1989). *Laws and Symmetry*. Oxford University Press, New York.

Vineberg, S. (2016). Dutch book arguments. In Zalta, E. N. (ed.), *The Stanford Encyclopedia of Philosophy*. Metaphysics Research Lab, Stanford University, spring 2016 edition.

Weisberg, M. and Reisman, K. (2008). The robust Volterra principle. *Philosophy of Science*, 75(1):106–131.

Wheeler, G. and Williamson, J. (2011). Evidential probability and objective Bayesian epistemology. In Bandyopadhyay, P. S. and Forster, M. R. (eds.), *Handbook of the Philosophy of Science*, vol. 7: Philosophy of Statistics, pages 307–331. Elsevier Science & Technology.

Williamson, J. (2007). Motivating Objective Bayesianism: From Empirical Constraints to Objective Probabilities. In Harper, W. and Wheeler, G. (eds.), *Probability and Inference: Essays in Honour of Henry E. Kyburg Jr.*, pages 151–179. College Publications, London.

Williamson, J. (2010). *In Defence of Objective Bayesianism*. Oxford University Press, Oxford.

Williamson, J. (2011). Objective Bayesianism, Bayesian conditionalisation, and voluntarism. *Synthese*, 178(1):67–85.

Williamson, J. (2017). *Lectures on Inductive Logic*. Oxford University Press, Oxford.

Wimsatt, W. C. (1981). Robustness, reliability, and overdetermination. In Brewer, M. B. and Collins, B. E. (eds.), *Scientific Inquiry and the Social Sciences*, pages 125–163. Jossey-Bass. Page references are to the version reprinted in (Wimsatt, 2007).

Wimsatt, W. C. (1994). The ontology of complex systems: Levels of organization, perspectives, and causal thickets. *Canadian Journal of Philosophy*, 24(sup1):207–274. Page references are to the version reprinted in (Wimsatt, 2007).

Wimsatt, W. C. (2007). *Re-Engineering Philosophy for Limited Beings*. Harvard University Press, Cambridge, MA.

Wimsatt, W. C. (2011). Robust re-engineering: a philosophical account? *Biology & Philosophy*, 26(2):295–303.

Winsberg, E. (2018). *Philosophy and Climate Science*. Cambridge University Press, Cambridge.

Wittgenstein, L. (1922). *Tractatus Logico-Philosophicus*. Routledge, Abingdon, 1974 edition. Translated by D. F. Pears and B. F. McGuiness.

Woodward, J. (2006). Some varieties of robustness. *Journal of Economic Methodology*, 13(2):219–240.

Cambridge Elements ☰

Philosophy of Science

Jacob Stegenga
University of Cambridge

Jacob Stegenga is a Reader in the Department of History and Philosophy of Science at the University of Cambridge. He has published widely on fundamental topics in reasoning and rationality and philosophical problems in medicine and biology. Prior to joining Cambridge he taught in the United States and Canada, and he received his PhD from the University of California San Diego.

About the Series

This series of Elements in Philosophy of Science provides an extensive overview of the themes, topics and debates which constitute the philosophy of science. Distinguished specialists provide an up-to-date summary of the results of current research on their topics, as well as offering their own take on those topics and drawing original conclusions.

Cambridge Elements ≡

Philosophy of Science